Finite Element Method Simulation of 3D Deformable Solids

Synthesis Lectures on Visual Computing
Computer Graphics, Animation, Computational Photography, and Imaging

Editor
Brian A. Barsky, *University of California, Berkeley*

This series presents lectures on research and development in visual computing for an audience of professional developers, researchers and advanced students. Topics of interest include computational photography, animation, visualization, special effects, game design, image techniques, computational geometry, modeling, rendering, and others of interest to the visual computing system developer or researcher.

Finite Element Method Simulation of 3D Deformable Solids
Eftychios Sifakis and Jernej Barbič
2015

Virtual Crowds: Steps Toward Behavioral Realism
Mubbasir Kapadia, Nuria Pelechano, Jan Allbeck, and Norm Badler
2015

Efficient Quadrature Rules for Illumination Integrals: From Quasi Monte Carlo to Bayesian Monte Carlo
Ricardo Marques, Christian Bouville, Luís Paulo Santos, and Kadi Bouatouch
2015

Numerical Methods for Linear Complementarity Problems in Physics-Based Animation
Sarah Niebe and Kenny Erleben
2015

Mathematical Basics of Motion and Deformation in Computer Graphics
Ken Anjyo and Hiroyuki Ochiai
2014

GPU-Based Techniques for Global Illumination Effects
László Szirmay-Kalos, László Szécsi, and Mateu Sbert
2008

High Dynamic Range Image Reconstruction
Asla M. Sá, Paulo Cezar Carvalho, and Luiz Velho
2008

High Fidelity Haptic Rendering
Miguel A. Otaduy and Ming C. Lin
2006

A Blossoming Development of Splines
Stephen Mann
2006

Finite Element Method Simulation of 3D Deformable Solids

Eftychios Sifakis and Jernej Barbič

ISBN: 978-3-031-01457-4   paperback
ISBN: 978-3-031-02585-3   ebook

DOI 10.1007/978-3-031-02585-3

A Publication in the Springer series
SYNTHESIS LECTURES ON VISUAL COMPUTING: COMPUTER GRAPHICS, ANIMATION, COMPUTATIONAL PHOTOGRAPHY, AND IMAGING

Lecture #21
Series Editor: Brian A. Barsky, *University of California, Berkeley*
ISSN pending.

Finite Element Method Simulation of 3D Deformable Solids

Eftychios Sifakis
University of Wisconsin-Madison

Jernej Barbič
University of Southern California

SYNTHESIS LECTURES ON VISUAL COMPUTING: COMPUTER GRAPHICS, ANIMATION, COMPUTATIONAL PHOTOGRAPHY, AND IMAGING #21

ABSTRACT

This book serves as a practical guide to simulation of 3D deformable solids using the Finite Element Method (FEM). It reviews a number of topics related to the theory and implementation of FEM approaches: measures of deformation, constitutive laws of nonlinear materials, tetrahedral discretizations, and model reduction techniques for real-time simulation.

Simulations of deformable solids are important in many applications in computer graphics, including film special effects, computer games, and virtual surgery. The Finite Element Method has become a popular tool in many such applications. Variants of FEM catering to both offline and real-time simulation have had a mature presence in computer graphics literature. This book is designed for readers familiar with numerical simulation in computer graphics, who would like to obtain a cohesive picture of the various FEM simulation methods available, their strengths and weaknesses, and their applicability in various simulation scenarios. The book is also a practical implementation guide for the visual effects developer, offering a lean yet adequate synopsis of the underlying mathematical theory.

Chapter 1 introduces the quantitative descriptions used to capture the deformation of elastic solids, the concept of strain energy, and discusses how force and stress result as a response to deformation. Chapter 2 reviews a number of constitutive models, i.e., analytical laws linking deformation to the resulting force that has successfully been used in various graphics-oriented simulation tasks. Chapter 3 summarizes how deformation and force can be computed discretely on a tetrahedral mesh, and how an implicit integrator can be structured around this discretization. Finally, chapter 4 presents the state of the art in model reduction techniques for real-time FEM solid simulation and discusses which techniques are suitable for which applications. Topics discussed in this chapter include linear modal analysis, modal warping, subspace simulation, and domain decomposition.

KEYWORDS

Finite Element Method, large deformations, solid mechanics, model reduction

*In loving memory of our grandmothers
Mubeccel and Nina.*

Contents

CHAPTER 1

Elasticity in Three Dimensions

In this chapter, we focus on three-dimensional elastic bodies undergoing deformation in space and discuss how we can formulate quantitative descriptions for the deformed shape of an object and the forces resulting from it. To a certain extent, these formulations are analogous to similar concepts from mass-spring systems, or deformable elastic strands. For example, a Hookean spring would only need a record of its current length and orientation to unambiguously define its potential energy and elastic reaction force. An elastic strand, which in its rest configuration might be parameterized relative to arc-length as $\mathcal{C}_0(s)$, $s \in [0, L]$, would be fully described by a similar parametric curve $\mathcal{C}(s)$ in its deformed state if torsional forces are to be ignored. However, since a volumetric body is able to alter its shape in more complex ways than, for example, a one-dimensional elastic strand, many quantitative descriptors that may be familiar from simpler mechanical systems will need to be extended and become more expressive. For the time being, and until Chapter 3, we will not concern ourselves with discretization issues. Our discussion will focus on the continuous phenomenon of elastic deformation, as if we had infinite resolution at our disposal.

1.1 DEFORMATION MAP AND DEFORMATION GRADIENT

Our initial objective is to provide a concise mathematical description of the deformation that an elastic body has sustained. This formulation will lay the foundation for appropriate representations of other physical properties such as force and energy. We begin by placing the undeformed elastic object in a coordinate system, and denote by Ω the volumetric domain occupied by the object. This domain will be referred to as the *reference (or undeformed) configuration*, and we follow the convention that capital letters $\vec{X} \in \Omega$ are used when referring to individual material points in this undeformed shape. Note that the precise position and orientation of the undeformed elastic body within the reference space is not important and can be chosen at will, as long as the shape of the object corresponds to a rest configuration.

When the object undergoes deformation, every material point $\vec{X}$ is being displaced to a new *deformed* location as seen in Figure 1.1 (top) which is, by convention, denoted by a lowercase variable $\vec{x}$. The relationship between each material point and its respective deformed location is captured by the *deformation function* $\vec{\phi} : \mathbf{R}^3 \to \mathbf{R}^3$, which maps every material point $\vec{X}$ to its respective deformed location $\vec{x} = \vec{\phi}(\vec{X})$.

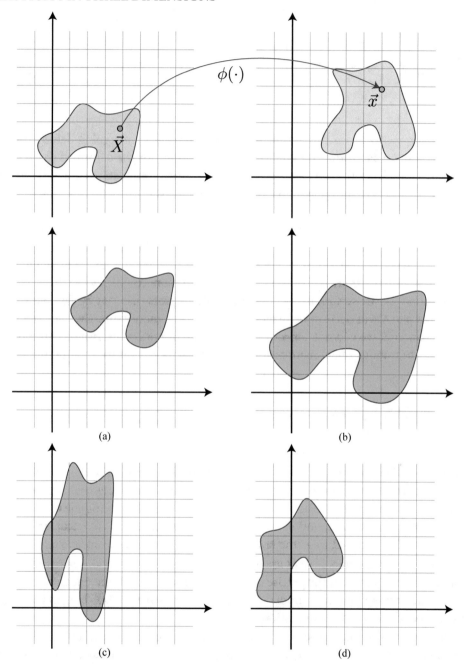

Figure 1.1: Top row: Illustration of the deformation map ϕ from the reference configuration (left) to the deformed shape (right). Bottom two rows: Sample deformations of (a) translation, (b) uniform scaling, (c) anisotropic scaling, and (d) rotation.

An important analytic quantity derived directly from $\vec{\phi}(\vec{X})$, whose utility will become apparent in the next sections, is the *deformation gradient* matrix $\mathbf{F} \in \mathbf{R}^{3\times 3}$. If we write $\vec{X} = (X_1, X_2, X_3)^T$ and $\vec{x} = \vec{\phi}(\vec{X}) = \left(\phi_1(\vec{X}), \phi_2(\vec{X}), \phi_3(\vec{X})\right)^T$ for the three components of the vector-valued function $\vec{\phi}$, the deformation gradient is written as:

$$\mathbf{F} := \frac{\partial(\phi_1, \phi_2, \phi_3)}{\partial(X_1, X_2, X_3)} = \left(\begin{array}{ccc} \partial\phi_1/\partial X_1 & \partial\phi_1/\partial X_2 & \partial\phi_1/\partial X_3 \\ \partial\phi_2/\partial X_1 & \partial\phi_2/\partial X_2 & \partial\phi_2/\partial X_3 \\ \partial\phi_3/\partial X_1 & \partial\phi_3/\partial X_2 & \partial\phi_3/\partial X_3 \end{array} \right)$$

or, in index notation $F_{ij} = \phi_{i,j} = \partial\phi_i/\partial X_j$ (indices given after the comma in the subscript indicate differentiation). That is, $\mathbf{F}$ is the Jacobian matrix of the deformation map. Although the academic deformation examples in Figure 1.1 correspond to scenarios where $\mathbf{F}$ is a constant, we note that in the general case, $\mathbf{F}$ will be spatially varying across Ω. In the next sections we will use the notation $\mathbf{F}(\vec{X})$ if such dependence needs to be made explicit.

Simple examples of deformation fields

The deformation depicted in the top row of Figure 1.1 is indicative of arbitrary shape changes that are likely to occur in animation tasks. For such instances of deformation it would not be possible to write $\phi(\vec{X})$ in closed form, and simulation would be employed instead to generate a numerical approximation. We can provide, however, some intuitive closed-form expressions for certain simple examples of deformation scenarios:

- Figure 1.1(a) depicts a configuration change that is merely a constant translation, say, by a vector $\vec{t}$. Here, the deformation map and gradient are:

$$\vec{x} = \phi(\vec{X}) = \vec{X} + \vec{t} \qquad \mathbf{F} = \partial\phi(\vec{X})/\partial\vec{X} = \mathbf{I}$$

- Figure 1.1(b) illustrates a scaling by a constant factor γ, specifically in our case a dilation by $\gamma = 1.5$. In this case, we have:

$$\phi(\vec{X}) = \gamma\vec{X} \qquad \mathbf{F} = \gamma\mathbf{I}$$

- In Figure 1.1(c) the reference shape has been scaled along the horizontal axis by a factor of 0.7, where the vertical axis is stretched by a factor of 2. Thus:

$$\left(\begin{array}{c} x \\ y \end{array} \right) = \phi(\vec{X}) = \phi\left(\begin{array}{c} X \\ Y \end{array} \right) = \left(\begin{array}{c} 0.7X \\ 2Y \end{array} \right) \qquad \mathbf{F} = \left(\begin{array}{cc} 0.7 & 0 \\ 0 & 2 \end{array} \right)$$

- The configuration of Figure 1.1(d) is the result of a 45° counter-clockwise rotation around the origin. Therefore, we have:

$$\begin{pmatrix} x \\ y \end{pmatrix} = \phi(\vec{X}) = \begin{pmatrix} \cos 45° & -\sin 45° \\ \sin 45° & \cos 45° \end{pmatrix} \begin{pmatrix} X \\ Y \end{pmatrix} \qquad \mathbf{F} = \begin{pmatrix} \cos 45° & -\sin 45° \\ \sin 45° & \cos 45° \end{pmatrix}$$

1.2 STRAIN ENERGY AND HYPERELASTICITY

One of the consequences of elastic deformation is the accumulation of potential energy in the deformed body, which is referred to as *strain energy* in the context of elastic deformable solids. We use the notation $E[\phi]$ for the strain energy, which suggests that the energy is fully determined by the deformation map of a given configuration. Although this is a quite broad definition, this statement nevertheless encapsulates a significant hypothesis that led to this formulation: we have assumed that the potential energy associated with a deformed configuration only depends on the final deformed shape, and not on the deformation path over time that brought the body into its current configuration. A similar subtle assumption is that the velocity of such a deformable solid would only contribute to the *kinetic* energy, while the potential energy denoted above as $E[\phi]$ would not be modulated by velocities in any way (this would not be an approproate description of the potential energy of *viscoelastic* materials). The independence of the strain energy on the prior deformation history is a characteristic property of so-called *hyperelastic* materials (which is the only class of materials we will address in this manuscript). This property is closely related with the fact that elastic forces of hyperelastic materials are *conservative* and *velocity-independent*: the total work done by the internal elastic forces in a deformation path depends solely on the initial and final configurations, not on the path itself. Note that any separately defined *damping* forces need not abide by this rule.

Different parts of a deforming body undergo shape changes of different severity. As a consequence, the relationship between deformation and strain energy is better defined on a local scale. We achieve that by introducing an *energy density* function $\Psi[\phi; \vec{X}]$, which measures the strain energy *per unit undeformed volume* on an infinitesimal domain dV around the material point $\vec{X}$. We can then obtain the total energy for the deforming body by integrating the energy density function over the entire domain Ω:

$$E[\phi] = \int_{\Omega} \Psi[\phi; \vec{X}] d\vec{X}.$$

Let us focus on a specific material location $\vec{X}_*$. Since the energy density $\Psi[\phi; \vec{X}_*]$ would only need to reflect the deformation behavior in an infinitesimal neighborhood of $\vec{X}_*$, we can reasonably approximate the deformation map in this tiny region using a first-order Taylor expansion:

$$\phi(\vec{X}) \approx \phi(\vec{X}_*) + \left.\frac{\partial \phi}{\partial \vec{X}}\right|_{\vec{X}_*} (\vec{X} - \vec{X}_*) = \vec{x}_* + \mathbf{F}(\vec{X}_*)(\vec{X} - \vec{X}_*)$$

$$= \underbrace{\mathbf{F}(\vec{X}_*)}_{\mathbf{F}_*} \vec{X} + \underbrace{\vec{x}_* - \mathbf{F}(\vec{X}_*)\vec{X}_*}_{\vec{t}} = \mathbf{F}_* \vec{X} + \vec{t}.$$

This equation suggests that, at least to a first-order approximation, $\Psi[\phi; \vec{X}_*]$ should be expressible as a function of $\mathbf{F}_*$ and $\vec{t}$, as these values fully parameterize the local Taylor expansion of ϕ near $\vec{X}_*$. Furthermore, we can expect that the value of the vector $\vec{t}$ would be irrelevant in this expression: different values of this parameter would indicate deformations that differ only by a constant translation, thus producing the same deformed shape and the same strain energy. Thus, we expect that the energy density function should be expressible as $\Psi[\phi; \vec{X}] = \Psi(\mathbf{F}(\vec{X}))$, i.e., a function of the local deformation gradient alone.

The previous arguments have simply established that the energy density function is expected to be a function of the deformation gradient. However, we have not provided specific formulas for $\Psi(\mathbf{F})$. This is intentional, as we want the flexibility to accommodate a variety of materials. Ultimately, the precise mathematical expression for $\Psi(\mathbf{F})$ will be the defining property of the material modeled.

What would a formula for $\Psi(\mathbf{F})$ look like?

Chapter 2 will provide concrete examples of material models and their associated energy definitions. For the time being, we list a few examples of (largely academic and oversimplified) hypothetical materials. A property we would naturally expect is that the energy should be bounded from below, making sure that minimum-energy states exist where the deforming object can settle to. For example:

$$\Psi(\mathbf{F}) = \frac{k}{2}\|\mathbf{F}\|_F^2 \qquad \text{where } k > 0.$$

This is an interesting hypothetical scenario. We would describe it as a "zero rest-volume material" in analogy to a "zero rest-length spring." The minimum energy is attained when $\mathbf{F} = \mathbf{0}$ throughout Ω, which means that $\phi(\vec{X}) = \text{const}$, i.e., all material points have the natural tendency to collapse down to a single point location.

Although such a material might be useful for "glueing" tasks, akin to zero rest-length springs, it is unnatural in the sense that the *reference* configuration Ω is not an *equilibrium* configuration. In order to preserve such a property, we would expect that the energy would exhibit a minimum at $\mathbf{F} = \mathbf{I}$, i.e., at the undeformed state $\phi(\vec{X}) = \vec{X}$. This could be achieved by setting:

$$\Psi(\mathbf{F}) = \frac{k}{2}\|\mathbf{F} - \mathbf{I}\|_F^2 \qquad \text{where } k > 0.$$

This model will have minimum energy when the object is in its reference configuration, or a constant translation away from it. Unfortunately, this model would not treat a *rotation* of the undeformed shape as a rest configuration, and the energy would be nonzero in this case. This lack of *rotational invariance* serves as motivation for the material models in the later sections of Chapter 2.

1.3   FORCE AND TRACTION

The next physical concept to be addressed is the elastic *force* incurred by a given deformation. First, consider a simple case from elementary mechanics: a small body (ideally dimensionless, i.e., a point mass) moving in a conservative force field. An easy example would be the gravitational field, where a body located at $\vec{x} = (x, y, z)$ has potential energy $E(\vec{x}) = m\vec{G} \cdot \vec{x} = mgz$, where $\vec{G} = (0, 0, g)$ with $g = 9.81 m/s^2$ is the acceleration of gravity (the z-axis is assumed to be vertical). We can then obtain the gravitational force as the negative gradient of the potential energy with respect to the object position $\vec{x}$:

$$\vec{f} = -\frac{\partial E(\vec{x})}{\partial \vec{x}} = (0, 0, -mg).$$

However, when attempting to express a similar relationship between force and energy for deformable bodies we need to be cautious of the fact that, in the absence of a prior discretization, such bodies form a continuous distribution of material, rather than a collection of isolated point masses. As a consequence, the appropriate quantitative description for elastic forces resulting from deformation would also be via a distribution. Thus, we use $\vec{f}(\vec{X})$ to denote force density, or more specifically *force-per-unit-undeformed volume*, in an infinitesimal region around $\vec{X}$. The aggregate force on a finite region $A \subset \Omega$ would then be computed by integrating

$$\vec{f}_{\text{aggregate}}(A) = \int_A \vec{f}(\vec{X}) d\vec{X}. \tag{1.1}$$

Unfortunately, this description is not appropriate for the force exerted by the body *along its boundary*. Consider an elastic body that is uniformly compressed (e.g., $\phi(\vec{X}) = \alpha\vec{X}$, $\alpha < 1$) to a lower volume. We would expect the body to react by pushing back against the apparatus that is causing the compression, and this restorative force would act along the surface of contact $S \subset \partial\Omega$. This time we define the *traction* $\vec{\tau}(\vec{X})$ to be the (surface) force density function that measures the *force-per-unit-undeformed area* along an infinitesimal region of the boundary surface $\vec{X}$. Once again, the aggregate force on a finite boundary region $B \subset \partial\Omega$ is computed by integrating:

$$\vec{f}_{\text{aggregate}}(B) = \oint_B \vec{\tau}(\vec{X}) dS. \tag{1.2}$$

Why treat (interior) force density and (boundary) traction separately? Ultimately, don't both of them just refer to standard elastic forces?

In loose terms, the reason is that on a force-per-volume basis, the net elastic force is quantitatively "stronger" on the boundary than on the interior; the force density would generally be a bounded function on the interior, but might look like a Dirac delta function on the boundary. This makes it possible to have a nonzero aggregate force along boundary patches, even though those would have had zero volume in an integral such as Equation (1.1). Instead of dealing with the peculiarities of integrating Delta functions just for the sake of having a single "force-per-unit-volume" descriptor, it makes better practical sense to separate force computation into the interior term of Equation (1.1) and the boundary term of Equation (1.2), where the integrands in either case are regular, finite-valued functions.

The question that remains is: how is it physically meaningful for elastic forces to have this apparent greater strength at the boundary? The important observation here is that $\vec{f}(\vec{X})$ is the total force that a point $\vec{X}$ receives from its surrounding material, *from all directions*. Although the force exceeded along each individual direction might be substantial, significant cancellation is to be expected when the force contributions of all directions are added up. For example, if we stretch a homogeneous material uniformly, each deformed material point will receive strong, yet equal (in magnitude) attractive forces along each direction, leading to a zero net force. Boundary points, on the other hand, only receive an elastic response from their material side, making it easier to accumulate a larger net force.

Finally, it is important to note that the distinction between force density and traction largely goes away once a discrete representation of the deformable body is adopted. In such case, we use *nodal forces* (instead of densities) as descriptors of the elastic material response, and their treatment is practically identical regardless of whether they reside on the boundary or interior of the deforming body.

1.4   THE FIRST PIOLA-KIRCHHOFF STRESS TENSOR

The differences between interior force density and boundary traction suggest that neither concept is fundamental enough to describe all aspects of the elastic response of deforming bodies. There is, however, a fundamental force descriptor that both such quantities can be derived from: the *stress tensor*. Intuitively, stress is a multidirectional extension to the notion of a force vector, which encapsulates information about both traction and force density, as we will see next.

There are a variety of "stress" descriptors that can be used for this purpose. For our discussion, we will focus on the *First Piola-Kirchhoff stress tensor* $\mathbf{P}$, a 3×3 matrix with the following properties:

- The internal traction at a boundary location $\vec{X} \in \partial\Omega$ is given by

$$\vec{\tau}(\vec{X}) = -\mathbf{P} \cdot \vec{N} \qquad (1.3)$$

where $\vec{N}$ is the outward pointing unit normal to the boundary *in the reference (undeformed) configuration*. This can serve as a formal definition for the stress tensor $\mathbf{P}$: for any interior point $\vec{X} \in \Omega \setminus \partial\Omega$ we can hypothetically slice the material with a cut through $\vec{X}$ and perpendicular to $\vec{N}$, and compute the traction along such a cut. Then, $\mathbf{P}$ would be the unique matrix that relates $\vec{\tau}$ and $\vec{N}$, as in Equation (1.3), for all possible boundary orientations.

- The internal force density can also be computed from $\mathbf{P}$, as follows:

$$\vec{f}(\vec{X}) = \mathbf{div}_{\vec{X}}\mathbf{P}(\vec{X}), \quad \text{or component-wise:} \quad f_i = \sum_{j=1}^{3} P_{ij,j} = \frac{\partial P_{i1}}{\partial X_1} + \frac{\partial P_{i2}}{\partial X_2} + \frac{\partial P_{i3}}{\partial X_3}.$$

We emphasize that the divergence operator and/or its component derivatives are taken with respect to the *undeformed/reference* coordinates $\vec{X}$.

- For hyperelastic materials, $\mathbf{P}$ is purely a function of the deformation gradient, and is related to the strain energy via the simple formula:

$$\mathbf{P}(\mathbf{F}) = \partial\Psi(\mathbf{F})/\partial\mathbf{F}.$$

As described, the First Piola-Kirchhoff stress tensor can be used to yield formulas both for force and tension, and is readily computed from the strain energy density definition. In fact, there are two equally popular (and, in fact, equivalent) ways to describe the material properties of a hyperelastic material: (a) an explicit formula for Ψ as a function of $\mathbf{F}$, or (b) an explicit formula for $\mathbf{P}$ as a function of $\mathbf{F}$. We will provide both types of definitions for all materials discussed in this document.

Example

In Section 1.2 we listed a hypothetical hyperelastic material with energy density $\Psi(F) = (k/2)\|\mathbf{F} - \mathbf{I}\|_F^2$. We are now in a position to give quantitative descriptions for the force and traction such a model would generate in response to deformation.

The Piola stress is computed as follows:

$$\delta\Psi(F) = (k/2)\delta\left[(\mathbf{F}-\mathbf{I}):(\mathbf{F}-\mathbf{I})\right] = k(\mathbf{F}-\mathbf{I}):\delta\mathbf{F} = \frac{\partial\Psi}{\partial\mathbf{F}}:\delta\mathbf{F},$$

thus $\mathbf{P} = \partial\Psi/\partial\mathbf{F} = k(\mathbf{F}-\mathbf{I})$, or component-wise: $P_{ij} = k(\phi_{i,j} - \delta_{ij})$.

From this, internal forces are computed as

$$f_i = \sum_j P_{ij,j} = \sum_j k\phi_{i,jj} = k\Delta\phi_i \Rightarrow \vec{f} = k\Delta\vec{\phi}.$$

Given such a material and appropriate boundary conditions, a rest configuration would be found by solving $\vec{f} = \vec{0}$ (in the absence of external forces) or $\Delta\vec{\phi} = \vec{0}$.

Lastly, let us assume a uniform expansion by a factor of 2. Thus $\phi(\vec{X}) = 2\vec{X}$, $\mathbf{F} = 2\mathbf{I}$, and $\mathbf{P} = k\mathbf{I}$. The traction that would result from this stress on a surface perpendicular to $\vec{N}$ would be $\vec{\tau} = -k\vec{N}$ (generating boundary forces that trigger inward motion toward the origin of the coordinate system, to restore the original shape and volume).

We note that, in much of the relevant engineering literature, a different notational convention is followed where $\vec{f}$ and $\vec{\tau}$ refer to the *externally applied* force density and traction, respectively. The relationship between these quantities and stress is then expressed by assuming that the body is in an equilibrium configuration, where such externally applied forces and tractions balance out exactly the internal elastic force and traction. The equations obtained under this convention would be:

$$\vec{f} + \mathbf{div}\,\mathbf{P} = 0, \quad \text{and} \quad \vec{\tau} = \mathbf{P} \cdot \vec{N}.$$

In this book, we will retain our original definition where $\vec{f}$ and $\vec{\tau}$ refer to internal forces, along with their respective relationships to $\mathbf{P}$ from earlier in this section, as these formulas hold true even if the deforming body is not in an equilibrium configuration. In cases where we need to refer to any externally applied force or traction we will use symbols $\vec{f}_{\text{ext}}$ and $\vec{\tau}_{\text{ext}}$, instead.

We provide a brief justification for the formulas relating the Piola stress $\mathbf{P}$ to force and traction. The intent of the derivations that follow is not to give a rigorous proof, but rather to explain the thought process that gave rise to these definitions.

Consider an arbitrary deformation $\vec{x} = \vec{\phi}(\vec{X})$, and a small perturbation $\delta\vec{\phi}(\vec{X})$ away from it. As the deforming body transitions from configuration $\vec{\phi}$ to the nearby configuration $\vec{\phi} + \delta\vec{\phi}$, the strain energy will be reduced by a certain amount δE equal to the work done by the elastic forces:

$$\delta E = -\int_{\Omega} \vec{f}(\vec{X}) \cdot \delta\vec{\phi}(\vec{X})d\vec{X} - \oint_{\partial\Omega} \vec{\tau}(\vec{X}) \cdot \delta\vec{\phi}(\vec{X})dS. \tag{1.4}$$

Note that the work is separately integrated in the interior and boundary regions, due to the quantitative difference of force and traction. The change in strain energy can also be expressed as:

$$\delta E = \delta\left[\int_{\Omega} \Psi(\mathbf{F})d\vec{X}\right] = \int_{\Omega} \delta\left[\Psi(\mathbf{F})\right]d\vec{X} = \int_{\Omega}\left[\frac{\partial\Psi}{\partial\mathbf{F}} : \delta\mathbf{F}\right]d\vec{X} = \int_{\Omega}[\mathbf{P} : \delta\mathbf{F}]\,d\vec{X}$$

$$= \sum_{i,j=1}^{3} \int_{\Omega} P_{ij} \delta F_{ij} d\vec{X} = \sum_{i,j=1}^{3} \int_{\Omega} P_{ij} \cdot \frac{\partial}{\partial X_j} \left[\delta \phi_i(\vec{X}) \right] d\vec{X}.$$

Using integration by parts, this is equivalently written as

$$\delta E = \sum_{i,j=1}^{3} \left[-\int_{\Omega} \frac{\partial}{\partial X_j} \left[P_{ij} \right] \cdot \delta \phi_i(\vec{X}) d\vec{X} + \oint_{\partial \Omega} P_{ij} N_j \cdot \delta \phi_i(\vec{X}) d\vec{X} \right]$$

$$= -\int_{\Omega} \mathbf{div} \mathbf{P} \cdot \delta \vec{\phi}(\vec{X}) d\vec{X} + \oint_{\partial \Omega} (\mathbf{P} \cdot \vec{N}) \cdot \delta \vec{\phi}(\vec{X}) d\vec{X}. \qquad (1.5)$$

From Equations (1.4, 1.5) and using the fundamental lemma of variational calculus we have that $\vec{f}(\vec{X}) = \mathbf{div} \mathbf{P}$, and $\vec{\tau}(\vec{X}) = -\mathbf{P}\vec{N}$.

CHAPTER 2

Constitutive Models of Materials

In this chapter we survey a number of different simulated materials and describe how their physical properties are encoded in their respective governing equations. The mathematical description of the physical traits of a given material is referred to as its *constitutive model* and includes the equations that relate stimuli (e.g., deformations) to the material responses (e.g., force, stress, energy) they trigger. In the spirit of the preceding chapter, two possibilities for what a *constitutive equation* can be are given by the formula for the Piola stress $\mathbf{P}$ as a function of the deformation gradient $\mathbf{F}$, or the formula for the energy density Ψ as a function of $\mathbf{F}$. For simplicity, we will focus on *isotropic* materials, whose response to deformation is independent of the orientation that such deformation is applied in.

2.1 STRAIN MEASURES

In principle, an explicit formula that relates Ψ and $\mathbf{F}$ (or $\mathbf{P}$ and $\mathbf{F}$) would be perfectly adequate as a constitutive equation: think of the formula $\Psi(\mathbf{F}) = \|\mathbf{F} - \mathbf{I}\|_F^2$ from the previous chapter as an example of this fact. The challenge, however, with designing constitutive models in this fashion is that using the raw elements of the matrix $\mathbf{F}$ can be a very unintuitive way to argue about the flavor and severity of a given deformation. Perhaps a certain material's response is dominated by its affinity for volume conservation, while a different material might prioritize resistance to shear. One would imagine that metrics such as "the ratio of volumetric expansion" or "the shear angle" would be much more effective in expressing the severity of the types of deformation that are most relevant to such materials. As a consequence, it is common for the design process for constitutive models to define certain intermediate quantities (examples of which are *strain measures* and *invariants*, discussed in this chapter) that are derived from $\mathbf{F}$, yet capture the specific traits of the deformation that the energy or stress values depend on more concisely than the deformation gradient itself.

A *strain measure* is intended to be a quantitative descriptor for the severity of a given deformation, i.e., a way to gauge how far this configuration is from a *rest configuration*. For this reason, although strain measures are derived from the deformation gradient, they strive to retain as much information from it that is relevant to assessing deformation magnitude, while disregarding any information contained in it that is unrelated to shape change. Consider the *Green strain tensor* $\mathbf{E} \in \mathbf{R}^{3 \times 3}$, defined as:

$$E = \frac{1}{2}\left(F^T F - I\right).$$ (2.1)

The Green strain tensor exemplifies many of the properties that we would ask for in a strain measure. When the body is in its reference configuration, i.e., $\vec{\phi}(\vec{X}) = \vec{X}$, we have $F = I$ and thus $E = 0$. The Green strain would also be zero if the elastic body is merely rotated and translated from its reference position, without changing its shape; in such a case $\vec{\phi}(\vec{X}) = R\vec{X} + \vec{t}$ (where R is a rotation matrix), thus $F = R$ and $E = 0$ since $R^T R = I$.

More generally, even for non-rigid motions, the deformation gradient can be decomposed as $F = RS$ into the product of a rotation matrix R, and a symmetric factor S via the polar decomposition. As a 3D rotation, matrix R encapsulates 3 degrees of freedom, while the symmetric S has 6 independent degrees of freedom. Substituting the polar decomposition into Equation (2.1) we obtain:

$$E = \frac{1}{2}\left(S^2 - I\right).$$

Thus, the Green strain succeeds in discarding the rotational degrees of freedom, which have no bearing on the severity of deformation, and retains the stretch/shear information in the 6-DOF symmetric factor S. This is also accomplished without explicitly forming the polar decomposition.

The price one has to pay for the useful properties the Green strain offers is that the expression of Equation (2.1) is a nonlinear (quadratic) function of deformation. This increases the complexity of constitutive models that are constructed based on it and, as we will see next, will lead to discretizations with nodal forces being nonlinear functions of nodal positions. In an effort to remedy this, we construct a *linear* approximation of Equation (2.1) by forming a Taylor expansion around the undeformed configuration $F = I$:

$$E(F) \approx \underbrace{E(I)}_{=0} + \left.\frac{\partial E}{\partial F}\right|_{F=I} : (F - I).$$

The derivative $\partial E/\partial F$ is most conveniently defined via the differential δE:

$$\frac{\partial E}{\partial F} : \delta F = \delta E = \frac{1}{2}\left(\delta F^T F + F^T \delta F\right).$$

If we evaluate the above at $F = I$, and subsequently substitute $\delta F \leftarrow F - I$ we obtain:

$$\left.\frac{\partial E}{\partial F}\right|_{F=I} : (F - I) = \frac{1}{2}\left[(F - I)^T I + I^T (F - I)\right] = \frac{1}{2}\left(F + F^T\right) - I.$$

The matrix resulting from this linear approximation of $E(F)$ is denoted by ϵ, where:

$$\epsilon = \frac{1}{2}\left(F + F^T\right) - I,$$

and called the *small strain tensor*, or the *infinitesimal strain tensor*. Intuitively, this is an acceptable approximation of the Green strain, if we know *a priori* that the object we are modeling will not be deforming too much. This could be the case, for example, when modeling very stiff or near-rigid materials such as buildings, steel tools, or vehicle frames (unless of course we're crash-testing them!). Note that small deformation doesn't necessarily imply small force—stiff materials would not deform too much, but would react to minute deformations with large forces.

The small strain tensor will give rise to a computationally lightweight constitutive model called *linear elasticity*, described in the next section, and enable discretizations which have a linear mapping between nodal positions and nodal elastic forces. As expected, this convenience comes with a certain limitation: the small strain tensor can be considered a reliable measure of deformation for *small* motions only while pronounced artifacts will occur if used in a large deformation scenario.

2.2   LINEAR ELASTICITY

The simplest practical constitutive model is *linear elasticity*, defined in terms of the strain energy density as:

$$\Psi(\mathbf{F}) = \mu \boldsymbol{\epsilon} : \boldsymbol{\epsilon} + \frac{\lambda}{2} \mathrm{tr}^2(\boldsymbol{\epsilon}), \tag{2.2}$$

where $\boldsymbol{\epsilon}$ is the small strain tensor and μ, λ are the *Lamé coefficients*, which are related to the the material properties of *Young's modulus* k (a measure of stretch resistance) and *Poisson's ratio* v (a measure of incompressibility) as:

$$\mu = \frac{k}{2(1+v)} \qquad \lambda = \frac{kv}{(1+v)(1-2v)}.$$

The relationship between the Piola stress $\mathbf{P}$ and $\mathbf{F}$ can be derived as follows:

$$\delta \boldsymbol{\epsilon} = \frac{1}{2} \left(\delta \mathbf{F} + \delta \mathbf{F}^T \right) = \mathrm{Sym}\{\delta \mathbf{F}\},$$

where $\mathrm{Sym}\{\mathbf{F}\}$ denotes the symmetric part of matrix $\mathbf{F}$. Furthermore,

$$\boldsymbol{\epsilon} : \delta \boldsymbol{\epsilon} = \boldsymbol{\epsilon} : \mathrm{Sym}\{\delta \mathbf{F}\} = \boldsymbol{\epsilon} : \delta \mathbf{F} \qquad \mathrm{tr}(\delta \boldsymbol{\epsilon}) = \mathbf{I} : \mathrm{Sym}\{\delta \mathbf{F}\} = \mathbf{I} : \delta \mathbf{F}$$

where the last part of each equality was due to the symmetry of $\boldsymbol{\epsilon}$ and $\mathbf{I}$, and finally:

$$\delta \Psi = 2\mu \boldsymbol{\epsilon} : \delta \boldsymbol{\epsilon} + \lambda \mathrm{tr}(\boldsymbol{\epsilon}) \mathrm{tr}(\delta \boldsymbol{\epsilon}) = \underbrace{[2\mu \boldsymbol{\epsilon} + \lambda \mathrm{tr}(\boldsymbol{\epsilon})\mathbf{I}]}_{=\partial \Psi / \partial \mathbf{F}} : \delta \mathbf{F}.$$

Since $\mathbf{P} = \frac{\partial \Psi}{\partial \mathbf{F}}$, this last equation implies that

$$\mathbf{P} = 2\mu \boldsymbol{\epsilon} + \lambda \mathrm{tr}(\boldsymbol{\epsilon})\mathbf{I},$$

or, after one final substitution for ϵ (and a few algebraic reductions):

$$\mathbf{P}(\mathbf{F}) = \mu(\mathbf{F} + \mathbf{F}^T - 2\mathbf{I}) + \lambda \operatorname{tr}(\mathbf{F} - \mathbf{I})\mathbf{I}.$$

These expressions allow us to make the following observations:

- The stress $\mathbf{P}$ is a *linear* function of the deformation gradient. As we will see in Chapter 3, this would also result on the nodal elastic forces having a linear dependence on nodal positions. As a consequence, this constitutive model is characterized by a significantly lower computational cost than other, nonlinear materials.

- Since the small strain tensor was designed to be accurate exclusively in a small deformation scenario, it would only be advisable to use linear elasticity when the magnitude of motion is small. For example, a rigid motion $\vec{\phi}(\vec{X}) = \mathbf{R}\vec{X} + \vec{t}$ would generally produce a non-zero strain $\epsilon = \frac{1}{2}(\mathbf{R} + \mathbf{R}^T) - \mathbf{I}$ and ultimately yield nonzero stress, even though no shape change has taken place.

The Partial Differential Equation (PDE) form of linear elasticity

For this simple material model, it is relatively straightforward to derive the differential equation that defines an equilibrium configuration. Assume an externally applied force distribution $\vec{f}_{\text{ext}}(\vec{X})$. When the object has settled to an equilibrium (rest) configuration, the deformation function will satisfy:

$$\mathbf{div}\,\mathbf{P} + \vec{f}_{\text{ext}} = 0 \Rightarrow \sum_{j=1}^{3} P_{ij,j} + f_{\text{ext}}^{(i)} = 0 \Rightarrow \qquad [\text{for } i = 1, 2, 3]$$

$$\Rightarrow -\sum_{j=1}^{3} \frac{\partial}{\partial X_j} \left[\mu(\phi_{i,j} + \phi_{j,i} - 2\delta_{ij}) + \delta_{i,j} \sum_{k=1}^{3} \lambda(\phi_{k,k} - 1) \right] = f_{\text{ext}}^{(i)} \Rightarrow$$

$$\Rightarrow -\sum_{j=1}^{3} \left[\mu(\phi_{i,jj} + \phi_{j,ij}) + \delta_{i,j} \sum_{k=1}^{3} \lambda \phi_{k,kj} \right] = f_{\text{ext}}^{(i)} \Rightarrow$$

$$\Rightarrow -\sum_{j=1}^{3} \left[\mu(\phi_{i,jj} + \phi_{j,ji}) \right] - \sum_{k=1}^{3} \lambda \phi_{k,ki} = f_{\text{ext}}^{(i)} \Rightarrow$$

(where we swapped the order, the partial derivatives were taken in $\phi_{j,ij} = \phi_{j,ji}$

and swapping summation variables j and k)

$$\Rightarrow -\sum_{j=1}^{3}\left[\mu\phi_{i,jj} + (\mu + \lambda)\phi_{j,ji}\right] = f_{\text{ext}}^{(i)} \Rightarrow$$

$$\Rightarrow -\mu\Delta\phi_i - (\mu + \lambda)\frac{\partial}{\partial X_i}[\nabla \cdot \vec{\phi}] = f_{\text{ext}}^{(i)} \Rightarrow$$

$$\Rightarrow -\mu\Delta\vec{\phi} - (\mu + \lambda)\nabla[\nabla \cdot \vec{\phi}] = \vec{f}_{\text{ext}}.$$

Which is a *linear*, second order Partial Differential Equation.

2.3   ST. VENANT-KIRCHHOFF MODEL

With the understanding that the small strain tensor is a mere approximation of the rotationally invariant Green strain $\mathbf{E}$, it makes sense to attempt an improvement of the linear elasticity model by using $\mathbf{E}$ in the place of ϵ in Equation (2.2):

$$\Psi(\mathbf{F}) = \mu\mathbf{E} : \mathbf{E} + \frac{\lambda}{2}\text{tr}^2(\mathbf{E}).$$

The resulting constitutive model is called a *St. Venant-Kirchhoff material*, and is the first truly nonlinear material we will examine. The first Piola-Kirchhoff stress tensor can be computed via a process similar to the one followed for linear elasticity:

$$\delta\mathbf{E} = \frac{1}{2}\left(\delta\mathbf{F}^T\mathbf{F} + \mathbf{F}^T\delta\mathbf{F}\right) = \text{Sym}\{\mathbf{F}^T\delta\mathbf{F}\},$$

and then, using the symmetry of $\mathbf{E}$ and $\mathbf{I}$:

$$\mathbf{E} : \delta\mathbf{E} = \mathbf{E} : \{\mathbf{F}^T\delta\mathbf{F}\} = \{\mathbf{F}\mathbf{E}\} : \delta\mathbf{F} \qquad \text{tr}(\delta\mathbf{E}) = \mathbf{I} : \{\mathbf{F}^T\delta\mathbf{F}\} = \mathbf{F} : \delta\mathbf{F}$$

$$\delta\Psi = 2\mu\mathbf{E} : \delta\mathbf{E} + \lambda\text{tr}(\mathbf{E})\text{tr}(\delta\mathbf{E}) = \underbrace{\mathbf{F}\left[2\mu\mathbf{E} + \lambda\text{tr}(\mathbf{E})\mathbf{I}\right]}_{=\partial\Psi/\partial\mathbf{F}} : \delta\mathbf{F}$$

$$\text{Thus} \quad \mathbf{P}(\mathbf{F}) = \mathbf{F}\left[2\mu\mathbf{E} + \lambda\text{tr}(\mathbf{E})\mathbf{I}\right]. \tag{2.3}$$

This is a rotationally invariant model; deformations that differ by a rigid body transformation are guaranteed to have the same strain energy. As a consequence, a St. Venant-Kirchhoff material exhibits plausible material response in many large deformation scenarios where linear elasticity would not be applicable. Equation (2.3) indicates that stress is a 3rd degree polynomial function of the components of $\mathbf{F}$; after discretization, nodal forces will likewise be expressed as cubic polynomials of nodal positions.

Although the St. Venant-Kirchhoff model offers significant benefits over a linear elastic model, its scope is limited to a certain degree due to its poor resistance to forceful compression: as a St. Venant-Kirchhoff elastic body is compressed, starting from its undeformed configuration, it reacts with a restorative force that initially grows with the degree of compression. However, once a critical compression threshold is reached ($\approx$ 58% of undeformed dimensions, when compression occurs along a single axis), the strength of the restorative force reaches a maximum. Further compression will be met with *decreasing* resistance, in fact the restorative force will vanish as the object is compressed all the way down to zero volume (an indication of this is that when $\mathbf{F} = \mathbf{0}$ we also have $\mathbf{P} = \mathbf{0}$). Continued compression past the point of zero volume (forcing the material to invert) will then create a restorative force that pushes the body toward complete inversion (reflection) along one or more axes. In practical simulation scenarios, this creates a material tendency to locally tangle and invert itself when subjected to strong compressive forces or constraints.

2.4 COROTATED LINEAR ELASTICITY

The use of the quadratic Green strain in the St. Venant-Kirchhoff materials guaranteed the rotational invariance of the constitutive model. At the same time, the increased complexity inherent in highly nonlinear materials leads to unintended side effects, such as the non-physical zero stress configurations of St. Venant-Kirchhoff materials under extreme compression. *Corotated linear elasticity* is a constitutive model that attempts to combine the simplicity of the stress-deformation relationship in a linear material with just enough nonlinear characteristics to secure rotational invariance.

Using the polar decomposition $\mathbf{F} = \mathbf{RS}$, we construct a new strain measure as $\boldsymbol{\epsilon}_c = \mathbf{S} - \mathbf{I}$, which is linear on the symmetric tensor $\mathbf{S}$ obtained by factoring away the rotational component of $\mathbf{F}$. Replacing the small strain tensor in Equation (2.2) we obtain the energy for corotational elasticity:

$$\Psi(\mathbf{F}) = \mu \boldsymbol{\epsilon}_c : \boldsymbol{\epsilon}_c + \frac{\lambda}{2}\text{tr}^2(\boldsymbol{\epsilon}_c) = \mu\|\mathbf{S} - \mathbf{I}\|_F^2 + (\lambda/2)\text{tr}^2(\mathbf{S} - \mathbf{I}),$$

which can also be equivalently written in any of the following ways:

$$\Psi(\mathbf{F}) = \mu\|\mathbf{F} - \mathbf{R}\|_F^2 + (\lambda/2)\text{tr}^2(\mathbf{R}^T\mathbf{F} - \mathbf{I})$$

$$\Psi(\mathbf{F}) = \mu\|\boldsymbol{\Sigma} - \mathbf{I}\|_F^2 + (\lambda/2)\text{tr}^2(\boldsymbol{\Sigma} - \mathbf{I}), \tag{2.4}$$

where $\boldsymbol{\Sigma}$ is the diagonal matrix with the singular values of $\mathbf{F}$, from the Singular Value Decomposition $\mathbf{F} = \mathbf{U}\boldsymbol{\Sigma}\mathbf{V}^T$. We can show that the First Piola-Kirchhoff stress tensor for corotated linear elasticity is given by:

$$\mathbf{P}(\mathbf{F}) = \mathbf{R}\left[2\mu\boldsymbol{\epsilon}_c + \lambda\text{tr}(\boldsymbol{\epsilon}_c)\mathbf{I}\right] = \mathbf{R}\left[2\mu(\mathbf{S} - \mathbf{I}) + \lambda\text{tr}(\mathbf{S} - \mathbf{I})\mathbf{I}\right]$$

$$= 2\mu(\mathbf{F} - \mathbf{R}) + \lambda\text{tr}(\mathbf{R}^T\mathbf{F} - \mathbf{I})\mathbf{R}. \tag{2.5}$$

Proof of the stress formula

Taking differentials of the Singular Value Decomposition $\mathbf{F} = \mathbf{U}\boldsymbol{\Sigma}\mathbf{V}^T$ we have:

$$\delta\mathbf{F} = (\delta\mathbf{U})\boldsymbol{\Sigma}\mathbf{V}^T + \mathbf{U}(\delta\boldsymbol{\Sigma})\mathbf{V}^T + \mathbf{U}\boldsymbol{\Sigma}\delta\mathbf{V}^T \Rightarrow$$

$$\Rightarrow \mathbf{U}^T(\delta\mathbf{F})\mathbf{V} = \underbrace{(\mathbf{U}^T\delta\mathbf{U})\boldsymbol{\Sigma}}_{(*)} + \delta\boldsymbol{\Sigma} + \underbrace{\boldsymbol{\Sigma}(\mathbf{V}^T\delta\mathbf{V})^T}_{(**)}. \tag{2.6}$$

For any orthogonal matrix $\mathbf{Q}$ we have

$$\mathbf{Q}^T\mathbf{Q} = \mathbf{I} \Rightarrow \delta(\mathbf{Q}^T\mathbf{Q}) = \mathbf{0} \Rightarrow (\delta\mathbf{Q})^T\mathbf{Q} + \mathbf{Q}^T\delta\mathbf{Q} = \mathbf{0} \Rightarrow (\mathbf{Q}^T\delta\mathbf{Q})^T = -\mathbf{Q}^T\delta\mathbf{Q}.$$

Thus, the matrices marked with $(*)$ and $(**)$ above are column- and row-scaled versions of skew symmetric matrices, and consequently have zero diagonal elements.

This implies that, if we restrict Equation (2.6) to its diagonal component only, terms $(*)$ and $(**)$ will vanish to yield the final expression for the differential of $\boldsymbol{\Sigma}$:

$$\delta\boldsymbol{\Sigma} = \mathrm{Diag}\{\mathbf{U}^T(\delta\mathbf{F})\mathbf{V}\}.$$

Using this result, the differential of Equation (2.4) becomes:

$$\delta\Psi = 2\mu(\boldsymbol{\Sigma} - \mathbf{I}) : \delta\boldsymbol{\Sigma} + \lambda\mathrm{tr}(\boldsymbol{\Sigma} - \mathbf{I})\mathrm{tr}(\delta\boldsymbol{\Sigma})$$

$$= 2\mu(\boldsymbol{\Sigma} - \mathbf{I}) : (\mathbf{U}^T\delta\mathbf{F}\mathbf{V}) + \lambda\mathrm{tr}\left[\mathbf{V}(\boldsymbol{\Sigma} - \mathbf{I})\mathbf{V}^T\right]\mathrm{tr}(\mathbf{U}^T\delta\mathbf{F}\mathbf{V})$$

$$= 2\mu\left[\mathbf{U}(\boldsymbol{\Sigma} - \mathbf{I})\mathbf{V}^T\right] : \delta\mathbf{F} + \lambda\mathrm{tr}(\mathbf{S} - \mathbf{I})\mathrm{tr}\left[(\mathbf{U}\mathbf{V}^T)^T\delta\mathbf{F}\right]$$

$$= 2\mu(\mathbf{F} - \mathbf{R}) : \delta\mathbf{F} + \lambda\mathrm{tr}(\mathbf{S} - \mathbf{I})\mathbf{R} : \delta\mathbf{F} = \mathbf{P} : \delta\mathbf{F},$$

from which Equation (2.5) follows.

The motivation behind corotational elasticity is to mimic what linear elasticity would have been, if the undeformed configuration had been rotated in the same way as encoded in the rotational factor $\mathbf{R}$ from the polar decomposition. Of course, in typical deformations where the value of $\mathbf{R}$ varies across the domain, making the transition from linear to corotated elasticity is more complex than a change of variables due to a (constant) rotation of the undeformed configuration. From a computational cost perspective, the overhead of corotated vs. linear elasticity includes the cost of the polar decomposition and the need to employ nonlinear solvers during simulation.

2.5   ISOTROPIC MATERIALS AND INVARIANTS

The constitutive models of St. Venant-Kirchoff material and Corotated linear elasticity have been constructed to be rotationally invariant. We can formally define this property by considering a pair of deformations, denoted by their deformation maps $\vec{\phi}_1(\vec{X})$ and $\vec{\phi}_2(\vec{X})$, that differ only by a rigid body transform, specifically:

$$\vec{\phi}_2(\vec{X}) = \mathbf{R}\vec{\phi}_1(\vec{X}) + \vec{t}, \quad \text{where } \mathbf{R} \text{ is a } 3 \times 3 \text{ rotation matrix.} \tag{2.7}$$

A constitutive model is rotationally invariant if and only if it guarantees that the strain energy will satisfy $E[\phi_1] = E[\phi_2]$ for any such deformation pair. For hyperelastic materials, an equivalent definition can be stated in terms of the strain energy density function. By taking gradients, we can see that any two deformations that satisfy Equation (2.7) will have deformation gradients related as $\mathbf{F}_2 = \mathbf{R}\mathbf{F}_1$. The energy density associated with these deformations must satisfy $\Psi(\mathbf{F}_1) = \Psi(\mathbf{F}_2)$, leading to the following equivalent definition of rotational invariance:

Definition: A hyperelastic constitutive model is *rotationally invariant* if and only if the strain energy density satisfies

$$\Psi(\mathbf{RF}) = \Psi(\mathbf{F})$$

for any value of the deformation gradient $\mathbf{F}$ and any 3×3 rotation matrix $\mathbf{R}$.

A consequence of this definition is that the strain energy in rotationally invariant models can be expressed solely as a function of the symmetric factor $\mathbf{S}$ from the polar decomposition of $\mathbf{F} = \mathbf{RS}$, since:

$$\Psi(\mathbf{F}) = \Psi(\mathbf{RS}) = \Psi(\mathbf{S}).$$

Although some model may define Ψ directly as a function of $\mathbf{S}$ (corotated elasticity was presented this way), we may avoid the need to compute the polar decomposition, if we are able to express Ψ as a function of some other intermediate quantity, which is a function of $\mathbf{S}$, yet also computable without an explicit polar decomposition. For example, St. Venant-Kirchhoff materials defined the energy density as a function of the Green strain $\mathbf{E} = \frac{1}{2}(\mathbf{S}^2 - \mathbf{I})$, which although fully determined by $\mathbf{S}$ can also be computed without an explicit polar decomposition as $\mathbf{E} = \frac{1}{2}(\mathbf{F}^T\mathbf{F} - \mathbf{I})$.

A similar, yet distinct property of certain constitutive models (including St. Venant-Kirchhoff and Corotated linear elasticity) is that of *isotropy*. In plain terms, a material is isotropic if its resistance to deformation is the same along all possible orientations that such deformation may be applied. Rubber and metal would be examples of isotropic materials, as they do not exhibit any particular direction/orientation along which are softer or stiffer. Steel-reinforced concrete would be an example of an anisotropic material, as its resistance to deformation is notably different along the direction of the steel supports, compared to a direction perpendicular to them. Human muscles are also quoted as an anisotropic structure, as a distinct material response is observed along the direction aligned with muscle fibers.

Isotropy is a property that needs to be assessed on a local scale, as it is always possible to generate directional features in larger structures by arranging material in specific ways (think of suspension bridges built from otherwise isotropic steel). In terms of a quantitative criterion for isotropy, we can think of an infinitesimal *spherical* volume of material dV, and consider the strain energy resulting from a prescribed deformation. Now, consider the scenario where we first transform the sphere dV by *rotating it about its center* and then apply the same deformation. If the material is isotropic, both scenarios would lead to the same strain energy. This is concretely expressed using the strain energy function, as defined below:

Definition: A hyperelastic constitutive model is *isotropic* if and only if the strain energy density satisfies

$$\Psi(\mathbf{F}\mathbf{Q}) = \Psi(\mathbf{F})$$

for any value of the deformation gradient $\mathbf{F}$ and any 3×3 rotation matrix $\mathbf{Q}$. A material that is both rotationally invariant and isotropic would satisfy

$$\Psi(\mathbf{R}\mathbf{F}\mathbf{Q}) = \Psi(\mathbf{F})$$

for arbitrary rotations $\mathbf{R}$ and $\mathbf{Q}$.

Using the Singular Value Decomposition $\mathbf{F} = \mathbf{U}\boldsymbol{\Sigma}\mathbf{V}^T$ we conclude that rotationally invariant, isotropic materials satisfy:

$$\Psi(\mathbf{F}) = \Psi(\mathbf{U}\boldsymbol{\Sigma}\mathbf{V}^T) = \dot{\Psi}(\boldsymbol{\Sigma}).$$

While the strain energy for rotationally invariant materials was a function of only 6 out of 9 degrees of freedom in $\mathbf{F}$ (those captured in the symmetric $\mathbf{S}$), for materials that are also isotropic, the energy density is actually only a function of the three singular values of $\mathbf{F}$. Equation (2.4) reveals that this is certainly the case for corotated linear elasticity. St. Venant-Kirchhoff can also be shown to satisfy all criteria for isotropy, after some simple algebraic manipulations. An example of a material that is rotationally invariant but *not* isotropic is described by the energy:

$$\Psi(\mathbf{F}) = \frac{k}{2}\vec{w}^T\mathbf{F}^T\mathbf{F}\vec{w},$$

where $\vec{w}$ is a given constant vector. This material behaves like a zero-restlength spring along the direction $\vec{w}$, while it does not have any resistance to deformation along directions perpendicular to $\vec{w}$ (notably, such energy formulas are often used to describe the behavior of anisotropic, fiber-laden structures like muscles).

Although it is possible to define an isotropic material by a relationship between Ψ and Σ (which encodes the only three relevant degrees of freedom in $\mathbf{F}$), this is not necessarily the preferred approach, since the overhead of an SVD computation would be necessary when evaluating any of these quantities. St. Venant-Kirchhoff materials avoided the need for an explicit polar decomposition, by using the Green strain E to convey (qualitatively) the same information as $\mathbf{S}$, while using a computationally inexpensive formula. For isotropic materials, this information is conveyed by the three *isotropic invariants* of the deformation gradient, which are equally expressive as the singular values, but can be computed inexpensively. Invariants are denoted by I_1, I_2, I_3 (or $I_1(\mathbf{F})$, etc., to emphasize the dependence on $\mathbf{F}$) and defined as:

$$I_1(\mathbf{F}) = \mathrm{tr}(\mathbf{F}^T\mathbf{F}), \qquad I_2(\mathbf{F}) = \mathrm{tr}\left[(\mathbf{F}^T\mathbf{F})^2\right], \qquad I_3(\mathbf{F}) = \det(\mathbf{F}^T\mathbf{F}) = (\det\mathbf{F})^2.$$

Their relationship to Σ is revealed by replacing $\mathbf{F}$ with its SVD in the previous expressions, where (after extensive cancellation) we obtain:

$$I_1 = \mathrm{tr}(\Sigma^2) = \sum_{i=1}^{3}\sigma_i^2, \qquad I_2 = \mathrm{tr}(\Sigma^4) = \sum_{i=1}^{3}\sigma_i^4, \qquad I_3 = \det(\Sigma^2) = \prod_{i=1}^{3}\sigma_i^2.$$

Also of use are the *derivatives* of the invariants with respect to the $\mathbf{F}$:

$$\delta I_1 = \delta[\mathrm{tr}(\mathbf{F}^T\mathbf{F})] = 2\mathrm{tr}(\mathbf{F}^T\delta\mathbf{F}) = (2\mathbf{F}):\delta\mathbf{F} \quad \Rightarrow \quad \frac{\partial I_1}{\partial\mathbf{F}} = 2\mathbf{F},$$

$$\delta I_2 = \delta[\mathrm{tr}(\mathbf{F}^T\mathbf{F}\mathbf{F}^T\mathbf{F})] = 4\mathrm{tr}(\mathbf{F}^T\mathbf{F}\mathbf{F}^T\delta\mathbf{F}) = (4\mathbf{F}\mathbf{F}^T\mathbf{F}):\delta\mathbf{F} \quad \Rightarrow \quad \frac{\partial I_2}{\partial\mathbf{F}} = 4\mathbf{F}\mathbf{F}^T\mathbf{F},$$

$$\delta I_3 = \delta[(\det\mathbf{F})^2] = 2\det\mathbf{F}\cdot\delta[\det\mathbf{F}] = 2(\det\mathbf{F})^2\mathbf{F}^{-T}:\delta\mathbf{F} \quad \Rightarrow \quad \frac{\partial I_3}{\partial\mathbf{F}} = 2I_3\mathbf{F}^{-T}.$$

When the common practice of defining an isotropic constitive model via invariants is followed, the strain energy density is provided as a function $\Psi(I_1, I_2, I_3)$. In such case, we can use the chain rule to compute the stress $\mathbf{P}$ as:

$$\mathbf{P} = \frac{\partial\Psi(I_1, I_2, I_3)}{\partial\mathbf{F}} = \frac{\partial\Psi}{\partial I_1}\frac{\partial I_1}{\partial\mathbf{F}} + \frac{\partial\Psi}{\partial I_2}\frac{\partial I_2}{\partial\mathbf{F}} + \frac{\partial\Psi}{\partial I_3}\frac{\partial I_3}{\partial\mathbf{F}}, \qquad \text{or, after substitution:}$$

$$\mathbf{P}(\mathbf{F}) = \frac{\partial\Psi}{\partial I_1}\cdot 2\mathbf{F} + \frac{\partial\Psi}{\partial I_2}\cdot 4\mathbf{F}\mathbf{F}^T\mathbf{F} + \frac{\partial\Psi}{\partial I_3}\cdot 2I_3\mathbf{F}^{-T} \qquad (2.8)$$

Finally, we note the additional invariant $J = \det\mathbf{F} = \sqrt{I_3}$ that is often used in replacement of I_3 while defining certain constitutive models. This quantity has an important physical interpretation as it represents the *fraction of volume change* due to deformation: a value of $J = 1$ implies that volume is preserved exactly, while $J = 2$ would indicate an expansion to twice the undeformed volume and $J = 0.2$ would be a compression down to 20% of the rest volume.

2.6   NEOHOOKEAN ELASTICITY

An example of an isotropic constitutive model defined via isotropic invariants is *Neohookean elasticity*:

$$\Psi(I_1, J) = \frac{\mu}{2}(I_1 - 3) - \mu \log(J) + \frac{\lambda}{2} \log^2(J), \quad \text{or equivalently}$$

$$\Psi(I_1, I_3) = \frac{\mu}{2}(I_1 - \log(I_3) - 3) + \frac{\lambda}{8} \log^2(I_3).$$

From this definition, we can easily compute

$$\frac{\partial \Psi}{\partial I_1} = \frac{\mu}{2} \quad \text{and} \quad \frac{\partial \Psi}{\partial I_3} = -\frac{\mu}{2I_3} + \frac{\lambda \log(I_3)}{4I_3}.$$

Thus, using Equation (2.8) we obtain:

$$\mathbf{P}(\mathbf{F}) = \mu \mathbf{F} - \mu \mathbf{F}^{-T} + \frac{\lambda \log(I_3)}{2} \mathbf{F}^{-T}$$

$$\text{or} \quad \mathbf{P}(\mathbf{F}) = \mu(\mathbf{F} - \mu \mathbf{F}^{-T}) + \lambda \log(J) \mathbf{F}^{-T}.$$

The Neohookean model has the following notable characteristics:

- By construction, the material exhibits a very strong reaction to extreme compression. Due to the logarithmic term $\log^2(J)$ in the energy, as $J \to 0$ we have $\Psi \to \infty$. This constructs a powerful energy barrier that strongly resists extreme compression. This is the only constitutive model we have seen so far that has this property. Models discussed earlier in this chapter will allow the material to compress to zero volume, even invert, while only absorbing a finite amount of energy.

- Modeling materials as strongly incompressible amounts to using a very large value for the second Lamé coefficient (λ). Doing so in the case of Neohookean elasticity would emphasize the $\log^2(J)$ energy term, and strongly enforce $J = 1$, which produces a volume-preserving formulation. Incidentally, setting a high value for λ in the earlier constitutive models does not quite have the desired effect, as their respective terms scaled by λ do not correspond to true volume change (as J does). For example, a high λ value for linear elasticity would enforce

$$\text{tr}(\mathbf{F} - \mathbf{I}) = 0 \Rightarrow \text{div}\left[\vec{\phi}(\vec{X}) - \vec{X}\right] = 0,$$

i.e., this will ensure that the displacement field $\vec{x}(\vec{X}) - \vec{X}$ is divergence free. This condition approximates volume preservation only for small deformations.

- The fact that the strain energy defines a (theoretically) impassable barrier at compression magnitudes leading to zero volume $J = 0$ implies that there is no mechanism for handling what happens when, accidentally, the simulated model is forced into a (theoretically impossible) inverted configuration. In such cases, energy and stress are undefined, since $J < 0$. We note that such inversions (although theoretically impossible) can easily occur in practice, as a result of nonphysical kinematic constraints, instability of time integration techniques, or inadequate convergence of numerical solvers. Should such a scenario arise, it is advised that the deformation gradient $\mathbf{F}$ be temporarily replaced by the nearest *physically plausible* value $\tilde{\mathbf{F}}$ (with $\det \tilde{\mathbf{F}} > \epsilon$).

CHAPTER 3

Discretization and Time Integration

The preceding chapters detailed a variety of physical laws that may be used to describe the response of elastic materials to deformation. Up to this point, these laws were expressed relative to a continuous deformation in space. Naturally, in order to enable numerical simulation all such laws have to be discretized; physical quantities such as the deformation map, the elastic strain energy, stress tensors, and elastic forces all have to be reformulated as functions of our discrete state variables.

3.1 ENERGY AND FORCE DISCRETIZATION

When modeling a deformable body on the computer we only store the values of the deformation map $\phi(\vec{X})$ on a finite number of points $\vec{X}_1, \vec{X}_2, \ldots, \vec{X}_N$, corresponding to the vertices of a discretization mesh. The respective deformed vertex locations $\vec{x}_i = \phi(\vec{X}_i)$, $i = 1, 2, \ldots, N$ are our discrete degrees of freedom, and we can write $\mathbf{x} = (\vec{x}_1, \vec{x}_2, \ldots, \vec{x}_N)$ for the aggregate state of our model. As a first step, we need to specify a method for reconstructing a continuous deformation map $\hat{\phi}$ from the discrete samples $\vec{x}_i = \phi(\vec{X}_i)$. In essence, this is just a choice of an interpolation scheme. For example, if a tetrahedral mesh is used to describe the deforming body, barycentric interpolation will extend the nodal deformations to the entire interior of the mesh. Trilinear interpolation would be a natural choice for lattice discretizations. At any rate, we denote the interpolated deformation map by $\hat{\phi}(\vec{X}; \mathbf{x})$, which emphasizes that this interpolated deformation is dependent on the discrete state $\mathbf{x}$.

For a hyperelastic material, the strain energy of any given deformation $\phi(\vec{x})$ is computed by integrating the energy density Ψ over the entire body Ω:

$$E[\phi] := \int_\Omega \Psi(\mathbf{F}) d\vec{X}.$$

We can now define a discrete energy, expressed as a function of the degrees of freedom $\mathbf{x}$, by simply plugging the interpolated deformation $\hat{\phi}$ into the definition of the strain energy

$$E(\mathbf{x}) := E\left[\hat{\phi}(\vec{X}; \mathbf{x})\right] = \int_\Omega \Psi\left(\hat{\mathbf{F}}(\vec{X}; \mathbf{x})\right) d\vec{X}, \tag{3.1}$$

where $\hat{\mathbf{F}}(\vec{X};\mathbf{x}) := \partial\hat{\phi}(\vec{X};\mathbf{x})/\partial\vec{X}$ is the deformation gradient computed from the interpolated map $\hat{\phi}$. It is understandable that Equation (3.1) may appear quite cryptic at this point, since both the energy $\Psi(\mathbf{F})$ and the interpolated $\hat{\phi}$ are likely defined via complex formulas. In this chapter we will focus on the common discretization scheme using linear tetrahedral elements and explain how the energy in Equation (3.1) and all its derived quantities can be evaluated systematically and efficiently.

Having defined the discrete energy $E(\mathbf{x})$, we can now compute the elastic forces associated with individual mesh nodes by taking the negative gradient of the elastic energy with respect to the corresponding degree of freedom:

$$\vec{f}_i(\mathbf{x}) = -\frac{\partial E(\mathbf{x})}{\partial\vec{x}_i} \quad\text{or, collectively }\mathbf{f} := (\vec{f}_1, \vec{f}_2, \dots, \vec{f}_N) = -\frac{\partial E(\mathbf{x})}{\partial\mathbf{x}}.$$

Outline of a brief (albeit over-simplified) "proof"

For simplicity, let us assume that (a) the deforming body is not subject to any internal friction forces that would reduce its overall energy, and (b) the mass of the body is distributed exclusively to the mesh nodes. The total energy of the body is the sum of strain energy (E) and kinetic energy (K) as follows:

$$E_{\text{total}} = E(\mathbf{x}) + K(\mathbf{v}) = E(\mathbf{x}) + \sum_{i=1}^{N}\frac{1}{2}m_i\|\vec{v}_i\|^2.$$

Since no friction forces are in effect the total energy is conserved over time, thus:

$$\frac{\partial}{\partial t}E_{\text{total}} = 0 \Rightarrow \sum_{i=1}^{N}\left[\frac{\partial E(\mathbf{x})}{\partial\vec{x}_i}\cdot\vec{v}_i + m_i\vec{a}_i\cdot\vec{v}_i\right] = 0.$$

Since the last equality holds for any value of the velocities $\{\vec{v}_i\}$, we must have:

$$\frac{\partial E(\mathbf{x})}{\partial\vec{x}_i} + m_i\vec{a}_i = 0 \Rightarrow \vec{f}_i = m_i\vec{a}_i = -\frac{\partial E(\mathbf{x})}{\partial\vec{x}_i} \quad\text{for all }i = 1, 2, \dots, N.$$

In practice, prior to computing each force, we first separate the energy integral of Equation (3.1) into the contributions of individual elements Ω_e (e.g., triangles, hexahedra, etc.) as follows:

$$E(\mathbf{x}) = \sum_{e}E^e(\mathbf{x}) = \sum_{e}\int_{\Omega_e}\Psi\left(\hat{\mathbf{F}}(\vec{X};\mathbf{x})\right)d\vec{X}.$$

Subsequently, the force $\vec{f}_i$ on each node can be computed by adding the contributions of all elements in its immediate neighborhood $\mathcal{N}_i$:

$$\vec{f}_i(\mathbf{x}) = \sum_{e \in \mathcal{N}_i} \vec{f}_i^e(\mathbf{x}), \quad \text{where} \quad \vec{f}_i^e(\mathbf{x}) = -\frac{\partial E^e(\mathbf{x})}{\partial \vec{x}_i}.$$

For simplicity, the following sections will focus on computing the nodal forces on an element-by-element basis, with the understanding that the aggregate forces are computed by accumulating the contributions from all elements in the mesh.

3.2   LINEAR TETRAHEDRAL ELEMENTS

Tetrahedral meshes are among the most popular discrete volumetric geometry representations. At the same time, they offer one of the most straightforward options for constructing a discretization of the elasticity equations. The convenience of tetrahedral discretizations is largely due to the simple interpolation method they imply; the reconstructed deformation map $\hat{\phi}$ can be defined to be a *piecewise linear* function over each tetrahedron. Specifically, in every tetrahedron $\mathcal{T}_i$ we have

$$\hat{\phi}(\vec{X}) = \mathbf{A}_i \vec{X} + \vec{b}_i \quad \text{for all} \ \ \vec{X} \in \mathcal{T}_i, \tag{3.2}$$

where the matrix $\mathbf{A}_i \in \mathbf{R}^{3 \times 3}$ and the vector $\vec{b}_i \in \mathbf{R}^3$ are specific to each tetrahedron. The interpolation scheme implied by Equation (3.2) is no other than simple barycentric interpolation on every element. Differentiating (3.2) with respect to $\vec{X}$ reveals that the deformation gradient $\mathbf{F} = \partial \hat{\phi} / \partial \vec{X} = \mathbf{A}_i$ is constant on each element and, as a consequence, so will be any discrete strain measure and stress tensor; this justifies why linear tetrahedral elements are also referred to as *constant strain tetrahedra*.

For simplicity of notation we write

$$\phi(\vec{X}) = \mathbf{F}\vec{X} + \vec{b},$$

where we dropped the tetrahedron index and replaced matrix $\mathbf{A}_i$ with its equal deformation gradient. Interestingly, it is possible to determine $\mathbf{F}$ (and $\vec{b}$, if desired) directly from the locations of the tetrahedron vertices, without involving any reasoning related to barycentric interpolation. Let us denote with $\vec{X}_1, \ldots, \vec{X}_4$ the undeformed (reference) locations of the tetrahedron vertices, and let $\vec{x}_1, \ldots, \vec{x}_4$ symbolize the respective deformed vertex locations as illustrated in Figure 3.1. Each vertex must satisfy $\vec{x}_i = \phi(\vec{X}_i)$, or

$$\left.\begin{cases} \vec{x}_1 = \mathbf{F}\vec{X}_1 + \vec{b} \\ \vec{x}_2 = \mathbf{F}\vec{X}_2 + \vec{b} \\ \vec{x}_3 = \mathbf{F}\vec{X}_3 + \vec{b} \\ \vec{x}_4 = \mathbf{F}\vec{X}_4 + \vec{b} \end{cases}\right\} \Rightarrow \left.\begin{cases} \vec{x}_1 - \vec{x}_4 = \mathbf{F}\left(\vec{X}_1 - \vec{X}_4\right) \\ \vec{x}_2 - \vec{x}_4 = \mathbf{F}\left(\vec{X}_2 - \vec{X}_4\right) \\ \vec{x}_3 - \vec{x}_4 = \mathbf{F}\left(\vec{X}_3 - \vec{X}_4\right) \end{cases}\right\},$$

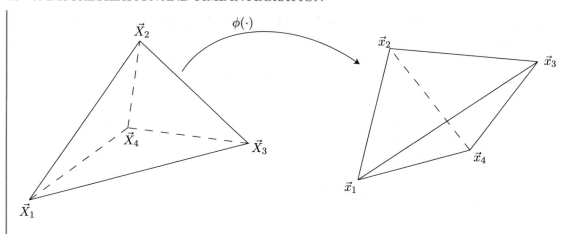

Figure 3.1: Reference (left) and deformed (right) shape of a linear tetrahedron.

where the last system was derived by subtracting the equation $\vec{x}_4 = \mathbf{F}\vec{X}_4 + \vec{b}$ from the three others, to eliminate the vector $\vec{b}$. It is possible to group the last three (vector) equations as a single matrix equation, by placing each one into the respective column of a 3×3 matrix:

$$
\begin{aligned}
\begin{bmatrix} \vec{x}_1 - \vec{x}_4 & \vec{x}_2 - \vec{x}_4 & \vec{x}_3 - \vec{x}_4 \end{bmatrix} &= \begin{bmatrix} \mathbf{F}\left(\vec{X}_1 - \vec{X}_4\right) & \mathbf{F}\left(\vec{X}_2 - \vec{X}_4\right) & \mathbf{F}\left(\vec{X}_3 - \vec{X}_4\right) \end{bmatrix} \\
\begin{bmatrix} \vec{x}_1 - \vec{x}_4 & \vec{x}_2 - \vec{x}_4 & \vec{x}_3 - \vec{x}_4 \end{bmatrix} &= \mathbf{F}\begin{bmatrix} \vec{X}_1 - \vec{X}_4 & \vec{X}_2 - \vec{X}_4 & \vec{X}_3 - \vec{X}_4 \end{bmatrix} \\
\mathbf{D}_s &= \mathbf{F}\mathbf{D}_m,
\end{aligned}
\tag{3.3}
$$

where

$$
\mathbf{D}_s := \begin{bmatrix} x_1 - x_4 & x_2 - x_4 & x_3 - x_4 \\ y_1 - y_4 & y_2 - y_4 & y_3 - y_4 \\ z_1 - z_4 & z_2 - z_4 & z_3 - z_4 \end{bmatrix}
\tag{3.4}
$$

is the *deformed shape matrix* and

$$
\mathbf{D}_m := \begin{bmatrix} X_1 - X_4 & X_2 - X_4 & X_3 - X_4 \\ Y_1 - Y_4 & Y_2 - Y_4 & Y_3 - Y_4 \\ Z_1 - Z_4 & Z_2 - Z_4 & Z_3 - Z_4 \end{bmatrix}
$$

is called the *reference shape matrix* (or "material-space" shape matrix).

We note that $\mathbf{D}_m$ is a constant matrix, as it only depends on the vertex coordinates in the reference (undeformed) configuration. Furthermore, the undeformed volume of the tetrahedron equals $W = \frac{1}{6}|\det \mathbf{D}_m|$, assuming that the reference shape of the tetrahedron is non-degenerate (i.e., nonzero volume, $W \neq 0$), the matrix $\mathbf{D}_m$ is nonsingular, and Equation (3.3) can be solved for $\mathbf{F}$ as:

$$\mathbf{F} = \mathbf{D}_s \mathbf{D}_m^{-1} \quad \text{or} \quad \mathbf{F}(\mathbf{x}) = \mathbf{D}_s(\mathbf{x})\mathbf{D}_m^{-1}, \tag{3.5}$$

where the last expression emphasizes that the deformed degrees of freedom appear only in the expression for $\mathbf{D}_s$, while the constant $\mathbf{D}_m^{-1}$ is precomputed and stored.

Since $\mathbf{F}$ is constant over the linear tetrahedron, the strain energy of this element reduces to:

$$E_i = \int_{T_i} \Psi(\mathbf{F})d\vec{X} = \Psi(\mathbf{F}_i) \int_{T_i} d\vec{X} = W \cdot \Psi(\mathbf{F}_i) \quad \text{or} \quad E(\mathbf{x}) = W \cdot \Psi(\mathbf{F}(\mathbf{x})). \tag{3.6}$$

We may subsequently use Equation (3.6) to derive the contribution of element T_i to the elastic forces on its four vertices as $\vec{f}_k^i = -\partial E_i(\mathbf{x})/\partial \vec{x}_k$. In fact, the forces on all four vertices can be collectively computed via the following equations:

$$\mathbf{H} = \begin{bmatrix} \vec{f}_1 & \vec{f}_2 & \vec{f}_3 \end{bmatrix} = -W\mathbf{P}(\mathbf{F})\mathbf{D}_m^{-T} \quad \text{and} \quad \vec{f}_4 = -\vec{f}_1 - \vec{f}_2 - \vec{f}_3, \tag{3.7}$$

where $\mathbf{P}(\mathbf{F})$ is the Piola stress as defined in Section 1.4.

Proof

Define $x_i^{(1)}, x_i^{(2)}, x_i^{(3)}$ to be the x-, y- and z- coordinates of the vertex $\vec{x}_i$. Likewise for the components of the nodal force $\vec{f}_i = (f_i^{(1)}, f_i^{(2)}, f_i^{(3)})$.

Lemma   For $i = 1, 2, 3$

$$\partial \mathbf{F}/\partial x_i^{(j)} = \mathbf{e}_j \mathbf{e}_i^T \mathbf{D}_m^{-1}.$$

Proof

From Equation (3.4) we have $\partial \mathbf{D}_s/\partial x_i^{(j)} = \mathbf{e}_j \mathbf{e}_i^T$. The Lemma follows directly from this equation and $\mathbf{F} = \mathbf{D}_s \mathbf{D}_m^{-1}$.

We proceed to compute the force component:

$$H_{ji} = f_i^{(j)} = -\frac{\partial E(\mathbf{x})}{\partial x_i^{(j)}} = -W \frac{\partial \Psi(\mathbf{x})}{\partial \mathbf{F}} : \frac{\partial \mathbf{F}}{\partial x_i^{(j)}} = -W\mathbf{P}(\mathbf{F}) : (\mathbf{e}_j \mathbf{e}_i^T \mathbf{D}_m^{-1}) =$$

$$= -W \operatorname{tr}\left[\mathbf{P}(\mathbf{F})\mathbf{D}_m^{-T}\mathbf{e}_i \mathbf{e}_j^T\right] = -W\mathbf{e}_j^T \mathbf{P}(\mathbf{F})\mathbf{D}_m^{-T}\mathbf{e}_i = \left[-W\mathbf{P}(\mathbf{F})\mathbf{D}_m^{-T}\right]_{ji}.$$

Thus $\mathbf{H} = -W\mathbf{P}(\mathbf{F})\mathbf{D}_m^{-T}$. The equation $\vec{f}_4 = -\vec{f}_1 - \vec{f}_2 - \vec{f}_3$ can be proved in a directly similar fashion, but is also a consequence of conservation of momentum; if the sum of all four nodal forces (which are internal to the body) did not sum to zero, this would violate conservation of linear momentum.

The computation of all elastic forces in a tetrahedral mesh is summarized in pseudocode as follows:

Algorithm 1 Batch computation of elastic forces on a tetrahedral mesh

---

Input : All the deformed vertex locations $\vec{x}_1, \vec{x}_2, \ldots, \vec{x}_N$ along with their reference (undeformed) counterparts $\vec{X}_1, \vec{X}_2, \ldots, \vec{X}_N$.
A tetrahedral mesh $\mathcal{M}$, listing each tetrahedron $\mathcal{T}_e = (i, j, k, l)$ via its four vertices.
Output (one-time precomputation) : A matrix $\mathbf{B}_m[e]$ and a scalar $W[e]$ for every element e.
Output (each force evaluation) : All the resulting elastic nodal forces $\vec{f}_1, \vec{f}_2, \ldots, \vec{f}_N$.

1: **for each** $\mathcal{T}_e = (i, j, k, l) \in \mathcal{M}$ **do**

2:    $\mathbf{D}_m \leftarrow \begin{bmatrix} X_i - X_l & X_j - X_l & X_k - X_l \\ Y_i - Y_l & Y_j - Y_l & Y_k - Y_l \\ Z_i - Z_l & Z_j - Z_l & Z_k - Z_l \end{bmatrix}$

3:    $\mathbf{B}_m[e] \leftarrow \mathbf{D}_m^{-1}$

4:    $W[e] \leftarrow \frac{1}{6} \det(\mathbf{D}_m)$ $\qquad\qquad\qquad$ ▷ W is the undeformed volume of $\mathcal{T}_e$

5: **end for** $\qquad\qquad\qquad$ ▷ Lines 1 through 5 can be factored away as precomputation

6: $\mathbf{f} \leftarrow \mathbf{0}$

7: **for each** $\mathcal{T}_e = (i, j, k, l) \in \mathcal{M}$ **do**

8:    $\mathbf{D}_s \leftarrow \begin{bmatrix} x_i - x_l & x_j - x_l & x_k - x_l \\ y_i - y_l & y_j - y_l & y_k - y_l \\ z_i - z_l & z_j - z_l & z_k - z_l \end{bmatrix}$

9:    $\mathbf{F} \leftarrow \mathbf{D}_s \mathbf{B}_m[e]$

10:    $\mathbf{P} \leftarrow \mathbf{P}(\mathbf{F})$ $\qquad\qquad\qquad\qquad\qquad\qquad\qquad\qquad$ ▷ From the constitutive law

11:    $\mathbf{H} \leftarrow -W[e]\mathbf{P}\,(\mathbf{B}_m[e])^T$

12:    $\vec{f}_i \mathrel{+}= \vec{h}_1, \vec{f}_j \mathrel{+}= \vec{h}_2, \vec{f}_k \mathrel{+}= \vec{h}_3$ $\qquad\qquad\qquad$ ▷ $\mathbf{H} = \begin{bmatrix} \vec{h}_1 & \vec{h}_2 & \vec{h}_3 \end{bmatrix}$

13:    $\vec{f}_l \mathrel{+}= (-\vec{h}_1 - \vec{h}_2 - \vec{h}_3)$

14: **end for**

3.3   FORCE DIFFERENTIALS

We have seen how discrete nodal forces $(\mathbf{f})$ can be computed for an arbitrary constitutive model, given nodal positions $(\mathbf{x})$ as input. This is all that is necessary to implement an explicit (e.g., Forward Euler) time-integration scheme; however, implicit methods such as Backward Euler will also require a process for computing *force differentials*, i.e., linearized nodal force increments around a configuration $\mathbf{x}_*$, relative to a small nodal force displacement $\delta\mathbf{x}$. We denote this by:

$$\delta\mathbf{f} = \delta\mathbf{f}(\mathbf{x}_*; \delta\mathbf{x}) := \left.\frac{\partial\mathbf{f}}{\partial\mathbf{x}}\right|_{\mathbf{x}=\mathbf{x}_*} \cdot \delta\mathbf{x}.$$

Although in this expression we used the *stiffness matrix* $\partial\mathbf{f}/\partial\mathbf{x}$ to aid in the definition of the force differential, in practice it may be preferable to avoid constructing this matrix explicitly, as the construction cost and memory footprint associated with it may impact performance. Instead, we aim to compute the force differentials $\delta\mathbf{f}$ directly, using only the information in the current state $\mathbf{x}_*$, the displacement $\delta\mathbf{x}$, and a small amount of additional meta-data.

As was the case with force computation, we evaluate the force differential vector $\delta\mathbf{f} = (\delta\vec{f}_1, \delta\vec{f}_2, \ldots, \delta\vec{f}_N)$ on an element-by-element basis, accumulating the contribution of each element to the aggregate value of each of its nodes. Consequently, we only focus on the process for computing differentials of nodal forces for a single tetrahedron. As before, we can pack the differentials of the first three vertices $(\delta\vec{f}_1, \delta\vec{f}_2$ and $\delta\vec{f}_3)$ in a single matrix representation:

$$\delta\mathbf{H} = \left[\delta\vec{f}_1 \; \delta\vec{f}_2 \; \delta\vec{f}_3\right].$$

Once $\delta\mathbf{H}$ has been evaluated, the force differential for the fourth node can be computed as $\delta\vec{f}_4 = -\delta\vec{f}_1 - \delta\vec{f}_2 - \delta\vec{f}_3$. Taking differentials on Equation (3.7), we obtain the following expression for $\delta\mathbf{H}$:

$$\delta\mathbf{H} = -W\delta\mathbf{P}(\mathbf{F}; \delta\mathbf{F})\mathbf{D}_m^{-T}.$$

Thus, the computation of nodal force differentials has been reduced to a computation of the stress differential $\delta\mathbf{F}$. There are two steps in completing this evaluation: (a) we need to construct the deformation gradient increment $\delta\mathbf{F}$ (the deformation gradient $\mathbf{F}$ itself is computed as detailed in the previous section) and (b) we need to provide a usable formula for $\delta\mathbf{P}(\mathbf{F}; \delta\mathbf{F})$.

We start with the differential of the deformation gradient $\delta\mathbf{F}$, which is easily computed by taking the differentials on Equation (3.5) to obtain:

$$\delta\mathbf{F} = (\delta\mathbf{D}_s)\mathbf{D}_m^{-1}.$$

Matrix $\delta\mathbf{D}_s$ itself is simply computed by arranging the nodal displacements in the same fashion as nodal positions were for $\mathbf{D}_s$:

$$\delta \mathbf{D}_s := \begin{bmatrix} \delta x_1 - \delta x_4 & \delta x_2 - \delta x_4 & \delta x_3 - \delta x_4 \\ \delta y_1 - \delta y_4 & \delta y_2 - \delta y_4 & \delta y_3 - \delta y_4 \\ \delta z_1 - \delta z_4 & \delta z_2 - \delta z_4 & \delta z_3 - \delta z_4 \end{bmatrix}.$$

The one remaining task is to provide a concise formula for $\delta \mathbf{P}(\mathbf{F}; \delta \mathbf{F})$. By necessity, this will be a process that depends on the constitutive model itself. Here, we provide examples of this derivation for the St. Venant-Kirchhoff and Neohookean material models:

Stress differentials for St. Venant-Kirchhoff materials

We start by assessing the differential of the Green strain tensor:

$$\mathbf{E} = \frac{1}{2}(\mathbf{F}^T \mathbf{F} - \mathbf{I}) \Rightarrow \delta \mathbf{E} = \frac{1}{2}(\delta \mathbf{F}^T \mathbf{F} + \mathbf{F}^T \delta \mathbf{F}).$$

We then proceed to compute the differential of the stress tensor itself:

$$\mathbf{P}(\mathbf{F}) = \mathbf{F} \left[2\mu \mathbf{E} + \lambda \mathrm{tr}(\mathbf{E})\mathbf{I} \right] \Rightarrow$$

$$\delta \mathbf{P}(\mathbf{F}; \delta \mathbf{F}) = \delta \mathbf{F} \left[2\mu \mathbf{E} + \lambda \mathrm{tr}(\mathbf{E})\mathbf{I} \right] + \mathbf{F} \left[2\mu \delta \mathbf{E} + \lambda \mathrm{tr}(\delta \mathbf{E})\mathbf{I} \right].$$

Stress differentials for Neohookean materials

We will use without proof the following two expressions for the differential of the matrix inverse and matrix determinant:

$$\delta[\mathbf{F}^{-1}] = -\mathbf{F}^{-1}\delta \mathbf{F}\mathbf{F}^{-1}, \text{ also } \delta[\mathbf{F}^{-T}] = -\mathbf{F}^{-T}\delta \mathbf{F}^T \mathbf{F}^{-T}$$

$$\delta[\det \mathbf{F}] = \det \mathbf{F} \cdot \mathrm{tr}(\mathbf{F}^{-1}\delta \mathbf{F}).$$

With these results, the differential of $\mathbf{P}$ is computed as:

$$\mathbf{P}(\mathbf{F}) = \mu(\mathbf{F} - \mathbf{F}^{-T}) + \lambda \log(J)\mathbf{F}^{-T} \Rightarrow$$

$$\Rightarrow \delta \mathbf{P}(\mathbf{F}; \delta \mathbf{F}) = \mu(\delta \mathbf{F} + \mathbf{F}^{-T}\delta \mathbf{F}^T \mathbf{F}^{-T}) + \lambda \frac{\delta[\det \mathbf{F}]}{J}\mathbf{F}^{-T} - \lambda \log(J)\mathbf{F}^T \delta \mathbf{F}^T \mathbf{F}^{-T}$$

$$\Rightarrow \delta \mathbf{P}(\mathbf{F}; \delta \mathbf{F}) = \mu \delta \mathbf{F} + \left[\mu - \lambda \log(J) \right] \mathbf{F}^{-T}\delta \mathbf{F}^T \mathbf{F}^{-T} + \lambda \mathrm{tr}(\mathbf{F}^{-1}\delta \mathbf{F})\mathbf{F}^{-T}.$$

The force differential computation is summarized in pseudocode as Algorithm (2).

Algorithm 2 Batch computation of elastic force differential on a tetrahedral mesh

---

Input : A tetrahedral mesh $\mathcal{M}$, listing each tetrahedron $\mathcal{T}_e = (i, j, k, l)$ via its four vertices.

Input : All the deformed vertex locations $\vec{x}_1, \vec{x}_2, \ldots, \vec{x}_N$ in the configuration relative to which the force differentials are computed.

Input : A complete set of vertex displacements $\delta\vec{x}_1, \delta\vec{x}_2, \ldots, \delta\vec{x}_N$ to be used in the force differential computation.

Prerequisites : We assume that the precomputation routine from algorithm 1 has been executed prior to this algorithm, and, after the last time, the vertex positions $(\vec{x}_1, \vec{x}_2, \ldots, \vec{x}_N)$ have changed. Thus, a matrix $\mathbf{B}_m[e]$ and a scalar $W[e]$ for every element e are available.

Output : All the resulting elastic nodal force differentials $\delta\vec{f}_1, \delta\vec{f}_2, \ldots, \delta\vec{f}_N$ resulting from the input displacements.

Optimization opportunity : If the routine below is expected to run several times in between updates to the vertex positions $(\vec{x}_1, \vec{x}_2, \ldots, \vec{x}_N)$, a deformation gradient $\mathbf{F}[e]$ can be precomputed (possibly as part of algorithm 1), circumventing the need to execute lines 2 and 4 of the algorithm below at each invocation.

1: **for each** $\mathcal{T}_e = (i, j, k, l) \in \mathcal{M}$ **do**

2: $\quad \mathbf{D}_s \leftarrow \begin{bmatrix} x_i - x_l & x_j - x_l & x_k - x_l \\ y_i - y_l & y_j - y_l & y_k - y_l \\ z_i - z_l & z_j - z_l & z_k - z_l \end{bmatrix}$

3: $\quad \delta\mathbf{D}_s \leftarrow \begin{bmatrix} \delta x_i - \delta x_l & \delta x_j - \delta x_l & \delta x_k - \delta x_l \\ \delta y_i - \delta y_l & \delta y_j - \delta y_l & \delta y_k - \delta y_l \\ \delta z_i - \delta z_l & \delta z_j - \delta z_l & \delta z_k - \delta z_l \end{bmatrix}$

4: $\quad \mathbf{F} \leftarrow \mathbf{D}_s \mathbf{B}_m[e]$

5: $\quad \delta\mathbf{F} \leftarrow (\delta\mathbf{D}_s)\mathbf{B}_m[e]$

6: $\quad \delta\mathbf{P} \leftarrow \delta\mathbf{P}(\mathbf{F}; \delta\mathbf{F})$ $\qquad\qquad\qquad\qquad\qquad$ ▷ From the stress derivative formula

7: $\quad \delta\mathbf{H} \leftarrow -W[e](\delta\mathbf{P})(\mathbf{B}_m[e])^T$

8: $\quad \delta\vec{f}_i \mathrel{+}= \delta\vec{h}_1, \delta\vec{f}_j \mathrel{+}= \delta\vec{h}_2, \delta\vec{f}_k \mathrel{+}= \delta\vec{h}_3$ $\qquad\qquad$ ▷ $\delta\mathbf{H} = \begin{bmatrix} \delta\vec{h}_1 & \delta\vec{h}_2 & \delta\vec{h}_3 \end{bmatrix}$

9: $\quad \delta\vec{f}_l \mathrel{+}= (-\delta\vec{h}_1 - \delta\vec{h}_2 - \delta\vec{h}_3)$

10: **end for**

3.4 AN IMPLICIT TIME INTEGRATION SCHEME

We are now in a position to describe a complete, implicit time integration scheme for nonlinear elastic bodies. The formulation that follows is based on the Backward Euler method, and thus is unconditionally stable for any timestep Δt (subject to the nonlinear equations involved being solved to satisfactory accuracy). We will first introduce some notation.

- $\mathbf{f}_e(\mathbf{x}^*)$: Elastic forces at configuration $\mathbf{x}^*$, as defined in previous sections.

- $\mathbf{K}(\mathbf{x}^*) = -\left.\frac{\partial \mathbf{f}_e}{\partial \mathbf{x}}\right|_{\mathbf{x}^*}$: This is the elasticity stiffness matrix evaluated around the configuration $\mathbf{x}^*$. In most cases, the matrix $\mathbf{K}$ will never be explicitly constructed; interative solvers that involve this matrix will only require the evaluation of matrix-vector products of the form $\mathbf{K}\mathbf{w}$. These products can be computed in a matrix-free fashion by calling the force differential computation procedure detailed in Algorithm (2) with an argument $\delta \mathbf{x} \leftarrow (-\mathbf{w})$.

- $\mathbf{f}_d(\mathbf{x}^*, \mathbf{v}^*) = -\gamma \mathbf{K}(\mathbf{x}^*)\mathbf{v}^*$: Damping forces at position $\mathbf{x}^*$ and velocity $\mathbf{v}^*$ according to the Rayleigh damping model. The parameter γ does not have a predetermined range (it is not confined to an interval such as $[0, 1]$) and can be spatially varying, or constant for simplicity.

- $\mathbf{f}(\mathbf{x}^*, \mathbf{v}^*) = \mathbf{f}_e(\mathbf{x}^*) + \mathbf{f}_d(\mathbf{x}^*, \mathbf{v}^*)$: The aggregate forces, including elastic and damping components.

- $\mathbf{M}$: The mass matrix. We shall assume $\mathbf{M}$ is lumped to diagonal form.

In order to define a backward Euler integration scheme, we will need to maintain both the position $(\mathbf{x}^n)$ and the velocity $(\mathbf{v}^n)$ of the deforming body at time t^n. Alternatively, it would have been possible to maintain just the two previous positions $\mathbf{x}^n$ and $\mathbf{x}^{n-1}$. The Backward Euler scheme computes the positions $\mathbf{x}^{n+1}$ and velocities $\mathbf{v}^{n+1}$ at time $t^{n+1}(:= t^n + \Delta t)$ as the solution of the (nonlinear) system of equations:

$$
\begin{aligned}
\mathbf{x}^{n+1} &= \mathbf{x}^n + \Delta t \mathbf{v}^{n+1} \qquad\qquad\qquad (3.8)\\
\mathbf{v}^{n+1} &= \mathbf{v}^n + \Delta t \mathbf{M}^{-1}\mathbf{f}(\mathbf{x}^{n+1}, \mathbf{v}^{n+1})\\
&= \mathbf{v}^n + \Delta t \mathbf{M}^{-1}\big(\mathbf{f}_e(\mathbf{x}^{n+1}) + \mathbf{f}_d(\mathbf{x}^{n+1}, \mathbf{v}^{n+1})\big). \qquad (3.9)
\end{aligned}
$$

Since the Backward Euler system is nonlinear due to Equation (3.9), we shall define an iterative process to compute the unknowns $\mathbf{x}^{n+1}$ and $\mathbf{v}^{n+1}$. We will construct sequences of approximations $\mathbf{x}^{n+1}_{(0)}, \mathbf{x}^{n+1}_{(1)}, \mathbf{x}^{n+1}_{(2)}, \ldots$ and $\mathbf{v}^{n+1}_{(0)}, \mathbf{v}^{n+1}_{(1)}, \mathbf{v}^{n+1}_{(2)}, \ldots$ respectively, such that $\mathbf{x}^{n+1}_{(k)} \overset{k\to\infty}{\longrightarrow} \mathbf{x}^{n+1}$ and $\mathbf{v}^{n+1}_{(k)} \overset{k\to\infty}{\longrightarrow} \mathbf{v}^{n+1}$ respectively. We will use the positions and velocities at the previous time step as initial guesses, i.e., $\mathbf{x}^{n+1}_{(0)} = \mathbf{x}^n, \mathbf{v}^{n+1}_{(0)} = \mathbf{v}^n$.

We introduce the position and velocity *correction* variables, defined as:

$$
\Delta \mathbf{x}_{(k)} := \mathbf{x}^{n+1}_{(k+1)} - \mathbf{x}^{n+1}_{(k)} \quad \text{and} \quad \Delta \mathbf{v}_{(k)} := \mathbf{v}^{n+1}_{(k+1)} - \mathbf{v}^{n+1}_{(k)}.
$$

In most cases, unless there is risk of ambiguity, we will drop the subscript and denote these corrections simply as $\Delta \mathbf{x}, \Delta \mathbf{v}$. At every step of our iterative scheme for the nonlinear Backward Euler system, we will linearize Equations (3.8) and (3.9) around the current iterate $\mathbf{x}_{(k)}^{n+1}, \mathbf{v}_{(k)}^{n+1}$, and the solution of the linearized system will define the next iterate $\mathbf{x}_{(k+1)}^{n+1}, \mathbf{v}_{(k+1)}^{n+1}$.

Lemma 3.1 $\Delta \mathbf{x} = \Delta t \, \Delta \mathbf{v}$.

Proof. Equation (3.8) is in fact linear. Therefore, at every iteration it will simply linearize to itself, i.e.,

$$\mathbf{x}_{(k)}^{n+1} = \mathbf{x}^n + \Delta t \mathbf{v}_{(k)}^{n+1}, \quad \text{for all } k.$$

Subtracting the above equations for iterations k and $k + 1$, we obtain

$$\mathbf{x}_{(k+1)}^{n+1} - \mathbf{x}_{(k)}^{n+1} = \Delta t (\mathbf{v}_{(k+1)}^{n+1} - \mathbf{v}_{(k)}^{n+1})$$

$$\text{or} \quad \Delta \mathbf{x} = \Delta t \, \Delta \mathbf{v}. \qquad \square$$

The linearization of Equation (3.9) around $(\mathbf{x}_{(k)}^{n+1}, \mathbf{v}_{(k)}^{n+1})$ yields:

$$
\begin{aligned}
\mathbf{v}_{(k)}^{n+1} + \Delta \mathbf{v} \;=\; & \mathbf{v}^n + \Delta t \mathbf{M}^{-1} \Big(\mathbf{f}_e(\mathbf{x}_{(k)}^{n+1}) + \left. \frac{\partial \mathbf{f}_e}{\partial \mathbf{x}} \right|_{\mathbf{x}_{(k)}^{n+1}} \cdot \Delta \mathbf{x} \\
& - \gamma \mathbf{K}(\mathbf{x}_{(k)}^{n+1})(\mathbf{v}_{(k)}^{n+1} + \Delta \mathbf{v}) \Big).
\end{aligned}
$$

Note that this equation is not quite an exact linearization, because in the damping term we fixed the stiffness matrix at the value it had around configuration $\mathbf{x}_{(k)}^{n+1}$ instead of performing a first-order Taylor expansion. This modification leads to a much simpler (modified) Newton scheme for the Backward Euler system and practically doesn't affect the convergence of the Newton scheme. We further manipulate the previous equation as follows:

$$
\begin{aligned}
\frac{1}{\Delta t^2} \mathbf{M} \Delta \mathbf{x} \;=\; & \frac{1}{\Delta t} \mathbf{M}(\mathbf{v}^n - \mathbf{v}_{(k)}^{n+1}) + \Big(\mathbf{f}_e(\mathbf{x}_{(k)}^{n+1}) \\
& - \mathbf{K}(\mathbf{x}_{(k)}^{n+1}) \Delta \mathbf{x} - \gamma \mathbf{K}(\mathbf{x}_{(k)}^{n+1})(\mathbf{v}_{(k)}^{n+1} + \frac{1}{\Delta t} \Delta \mathbf{x}) \Big)
\end{aligned}
$$

$$\left[\left(1+\frac{\gamma}{\Delta t}\right)\mathbf{K}(\mathbf{x}_{(k)}^{n+1})+\frac{1}{\Delta t^2}\mathbf{M}\right]\Delta\mathbf{x} =$$
$$= \frac{1}{\Delta t}\mathbf{M}(\mathbf{v}^n - \mathbf{v}_{(k)}^{n+1}) + \left(\mathbf{f}_e(\mathbf{x}_{(k)}^{n+1}) - \gamma\mathbf{K}(\mathbf{x}_{(k)}^{n+1})\mathbf{v}_{(k)}^{n+1}\right)$$
$$= \frac{1}{\Delta t}\mathbf{M}(\mathbf{v}^n - \mathbf{v}_{(k)}^{n+1}) + \left(\mathbf{f}_e(\mathbf{x}_{(k)}^{n+1}) + \mathbf{f}_d(\mathbf{x}_{(k)}^{n+1},\mathbf{v}_{(k)}^{n+1})\right)$$
$$= \frac{1}{\Delta t}\mathbf{M}(\mathbf{v}^n - \mathbf{v}_{(k)}^{n+1}) + \mathbf{f}(\mathbf{x}_{(k)}^{n+1},\mathbf{v}_{(k)}^{n+1}). \tag{3.10}$$

The system described by Equation (3.10) is symmetric and positive definite, and can be solved efficiently with a Krylov subspace method such as Conjugate Gradients. We also note that Equation (3.10) only determines the update for the positions at time t^{n+1}. Velocities should be updated at each iteration using the relation $\mathbf{v}_{(k+1)}^{n+1} = \mathbf{v}_{(k)}^{n+1} + \frac{1}{\Delta t}\Delta x$.

As a final observation, Equation (3.10) can be modified to yield a *quasistatic* simulation, where every configuration over time is the result of a rest configuration (subject to the imposed kinematic constraints and boundary conditions). We achieve this by setting $\Delta t \to \infty$, effectively indicating that at every simulated instance we allow infinite time for the elastic body to settle into an equilibrium configuration. The Newton iteration for this quasistatic problem simply becomes:

$$\mathbf{K}(\mathbf{x}_{(k)}^{n+1})\Delta\mathbf{x} = \mathbf{f}(\mathbf{x}_{(k)}^{n+1}),$$

after which positions are updated as

$$\mathbf{x}_{(k+1)}^{n+1} \leftarrow \mathbf{x}_{(k)}^{n+1} + \Delta\mathbf{x}.$$

CHAPTER 4

Model Reduction

4.1 INTRODUCTION

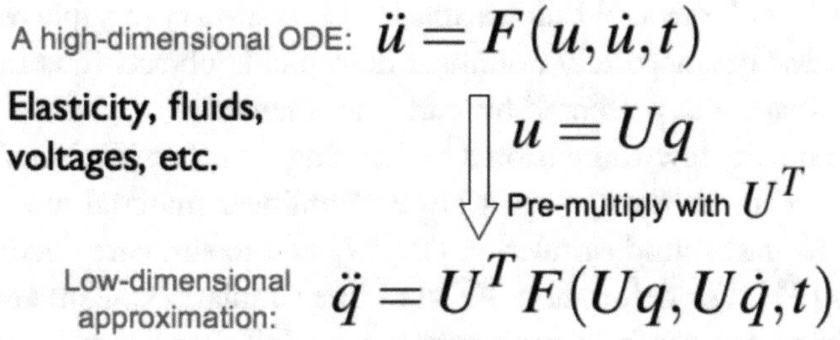

A high-dimensional ODE: $\ddot{u} = F(u, \dot{u}, t)$

Elasticity, fluids, voltages, etc.

$u = Uq$

Pre-multiply with U^T

Low-dimensional approximation: $\ddot{q} = U^T F(Uq, U\dot{q}, t)$

Figure 4.1: Model reduction overview: a high-dimensional ordinary differential equation is approximated with a projection to a low-dimensional space.

Model reduction (also-called dimensional model reduction, or model order reduction (MOR)) is a technique to simplify the simulation of dynamical systems described by differential equations. The idea is to project the original, high-dimensional, state-space onto a properly chosen low-dimensional subspace to arrive at a (much) smaller system having properties similar to the original system (see Figure 4.1). Complex systems can thus be approximated by simpler systems involving fewer equations and unknown variables, which can be solved much more quickly than the original problem. Such projection-based model reduction appears in literature under the names of *Principal Orthogonal Directions (POD) Method*, or *Subspace Integration Method*, and it has a long history in the engineering and applied mathematics literature [29]. See [27] and [33] for good overviews of model reduction applied to linear and nonlinear problems, respectively.

Model reduction has been used extensively in the fields of control theory, electrical circuit simulation, computational electromagnetics, and microelectromechanical systems [28]. Most model reduction techniques in these fields, however, aim at linear systems, and linear time-invariant systems in particular, e.g., small perturbations of voltages in some complex nonlinear circuit. Another common characteristic of these applications is that both the input and output are low-dimensional, i.e., one may want to study how the voltage level at some circuit location depends on the input voltage at another location, in a complex nonlinear circuit. In *computer graphics*, however, one is often interested in nonlinear systems (e.g., large deformations of objects) that exhibit interesting, very visible, dynamics. The output in computer graphics is usually high-dimensional, e.g., the deformation of an entire 3D solid object, or fluid velocities sampled on a high-resolution grid. For these reasons, many conventional reduction techniques do not immediately apply to computer-graphics problems.

4.1.1   SURVEY OF POD-BASED MODEL REDUCTION IN COMPUTER GRAPHICS

The initial model reduction applications for deformable object simulation in computer graphics investigated *linear* FEM deformable objects [14, 18, 32]. These models are very fast, but are (due to linear Cauchy strain) accurate only for small deformations and produce visible artifacts under large deformations. In order to avoid such artifacts, it is necessary to apply reductions to *nonlinear* elasticity. For *real-time* geometrically nonlinear deformable objects (quadratic Green-Lagrange strain), such an approach was presented by Barbič and James [4, 6], who also gave an automatic approach to select a quality low-dimensional basis, using *modal derivatives*. An and colleagues [2] demonstrated how to efficiently support arbitrary nonlinear material models. Model reduction has also been used for fast sound simulation [10, 17] and to simulate frictional contact between deformable objects [21]. For deformable FEM offline simulations, Kim and James [23] applied *online* model reduction to adaptively replace expensive full simulation-steps with reduced steps, which made it possible to *throttle* the simulation costs at run-time. Treuille and colleagues [39] applied model reduction to fluid simulation in computer graphics. Wicke and colleagues [41] improved Treuille's fluid method to support reduced fluid simulations on several (inter-connected) domains with specialized basis functions on each domain (domain decomposition for fluids). Recently, Barbič and Zhao [8] demonstrated a domain decomposition method for open-loop solid deformable models, by employing gradients of polar decomposition rotation matrices, whereas Kim and James [24] tackled a similar problem using inter-domain spring forces.

4.2   LINEAR MODAL ANALYSIS

Figure 4.2: Linear modes for a cantilever beam.

Although elastic objects can in principle deform arbitrarily, they tend to have a bias in deforming into certain characteristic, low-energy shapes. Most of us will remember the high-school physics example of a string stretched between two walls (or, say, a violin string), where one studies the natural frequencies ω_i of the string, together with their associated shapes, typically of the form $\sin(\omega_i x)$ and $\cos(\omega_i x)$. The same intuition carries over to arbitrary three-dimensional elastic objects deforming by a small amount around their rest configuration, whether it be a metal wire, a thin shell (e.g., cloth on a character), or a solid 3D tet mesh model of a skyscraper. The low-frequency *modes* are the deformations, which, for a given amount of displaced mass (or volume) subject to specific boundary conditions such as fixed vertices, increase the elastic strain energy of the object by the least amount. In other words, they are the shapes with the *least* resistance to deformation. How are the modes and frequencies computed? One has to first form the system mass and stiffness matrices $M \in \mathbb{R}^{3n \times 3n}$ and $K \in \mathbb{R}^{3n \times 3n}$, where n is the number of mesh vertices. The specific approach to compute M and K depends on each particular mechanical system. For example, the tet skyscraper may be modeled using 3D FEM elasticity, whereas for a cloth model one may compute M by lumping the mass at the vertices, and set K to the gradient of the internal cloth forces (in the rest configuration), computed, say, using the Baraff-Witkin cloth model [3]. Once M and K are known, one has to prescribe boundary conditions, i.e., specify how the object is constrained. The modes and frequencies greatly depend on this choice. It is possible to set no boundary conditions, in which case one obtains *free-flying modes*. In order to compute the modes, one forms matrices $\overline{M}$ and $\overline{K}$ where the rows and columns corresponding to the fixed degrees of freedom have been removed from M and K. Then, one solves the generalized eigenvalue problem

$$\overline{K}x = \lambda \overline{M}x. \tag{4.1}$$

Matrices $\overline{M}$ and $\overline{K}$ are typically large and sparse. One can solve the eigenvalue problem, say, using the Arnoldi iteration implemented by the ARPACK eigensolver [26]. This solver is free and has performed very well in various model reduction computer graphics projects by Jernej Barbič and other researchers in the field. Because $\overline{M}$ and $\overline{K}$ are symmetric positive-definite (in typical applications, e.g., deformable object in the rest configuration), the eigenvalues are real and non-negative. One seeks the smallest eigenvalues λ_i and their associated eigenvectors ψ_i, $i = 1, 2, \ldots, k$, where k is the number of modes to be retained. For objects with no constrained vertices (free-flying objects), the first six eigenvalues are zero and the modes correspond to rigid translations and infinitesimal rotations; these modes are typically discarded. The eigenvalues are squares of the natural frequencies of vibration, $\lambda_i = \omega_i^2$, and the eigenvectors ψ_i are the modes. It should be noted that one typically inserts zeros into ψ_i at locations of fixed degrees of freedom, so that the resulting vector is of length $3n$. In order to check that the eigensolver was successful, it is common to visualize the individual modes, by animating them as $\psi_i \sin(\omega_i t)$, where t is time. The different modal vectors are typically assembled into a *linear modal basis* matrix $U = [\psi_1, \ldots, \psi_k] \in \mathbb{R}^{3n \times k}$.

4.2.1 SMALL DEFORMATION SIMULATION USING LINEAR MODAL ANALYSIS

Small deformations $u \in \mathbb{R}^{3n}$ (n is the number of mesh vertices) follow the equation

$$M\ddot{u} + D\dot{u} + Ku = f, \tag{4.2}$$

where $M, D, K \in \mathbb{R}^{3n \times 3n}$ are the mass, damping, and stiffness matrices, respectively, and $f \in \mathbb{R}^{3n}$ are the external forces [35]. Equation 4.2 is a linear, high-dimensional ordinary differential equation, obtained by applying the Finite Element Method to the linearized partial differential equations of elasticity. It is most commonly applied to 3D solids, but can also model shells and strands. It is only accurate under small deformations; very visible artifacts appear under large deformations. Equation 4.2 models the object at full resolution (no reduction), which means that it incorporates transient effects such as (localized) waves traveling across the object. It is often employed, for example, to perform earthquake simulation, typically using supercomputers on large meshes involving millions of degrees of freedom [1]. For real-time applications, the computational costs of timestepping Equation 4.2 may be prohibitive. Instead, one can perform model reduction by approximating the deformation vector u as $u = Uz$, where $z \in \mathbb{R}^k$ is a vector of *modal amplitudes* (typically $k \ll 3n$). If we choose damping to be a linear combination of M and K, $D = \alpha M + \beta K$, for some scalars $\alpha, \beta > 0$ (this is called *Rayleigh damping*), and premultiply Equation 4.2 by U^T, Equation 4.2 projects to k *independent* one-dimensional ordinary differential equations

$$\ddot{z}_i + (\alpha + \beta \lambda_i)\dot{z}_i + \lambda_i z_i = \psi_i^T f, \tag{4.3}$$

for $i = 1, \ldots, k$. Here, we have used the fact that the modes are generalized eigenvectors $K\psi_i = \lambda_i M\psi_i$, and are therefore mass-orthonormal, $(M\psi_i)^T \psi_j = 0$ when $i \neq j$, and $(M\psi_i)^T \psi_i = 1$ for all i. The one-dimensional equations given in (4.3) can be timestepped independently. This can be done very efficiently (see, e.g., [18]). The full deformation can be reconstructed by multiplying $u = Uz$. This multiplication is fast when k is small (a few hundred modes). It can also be performed very efficiently in graphics hardware [18]. It can be shown that, as $k \to 3n$, this approximation converges to the solution of Equation 4.2. This property is very useful, as it makes it possible to trade computation accuracy for speed.

4.2.2 APPLICATION TO SOUND SIMULATION

Sound originates from the mechanical vibration of objects. These mechanical vibrations excite the surrounding medium (typically air or water), and the pressure waves then propagate to the listener location. The object mechanical vibrations are usually modeled using FEM and the equations of elasticity, whereas the pressure propagation is usually modeled by the wave equation. There are varying degrees of approximation that can be applied to each of these two tasks. Because deformation amplitudes in sound applications are small, linear elasticity is often employed for their simulation [17, 31]. However, richer, nonlinear sound can be produced using nonlinear simulation [10]. Linear simulation is straightforward and follows the material from Section 4.2.1. For each object that is to produce sound, one first has to set the fixed vertices (if any), and extract the modes. Next, one runs any physical simulation (typically rigid body simulation), producing contact forces. These forces are then used as external forces f for the modal oscillators in Equation 4.3, producing modal excitations $z_i(t)$. It remains to be described how $z_i(t)$ are used to generate the sound signal $s(t)$. A very common approach is to assign some meaningful weights

w_i to each mode, and compute sound as

$$s(t) = \sum_{i=1}^{k} w_i z_i(t), \qquad (4.4)$$

where k is the number of modes. The weights w_i can be set to a constant, $w_i = 1$ [31], or they can be made non-constant to model the fact that different modes radiate with different intensities. Alternatively, weights can be made to depend on the listener location x, $w_i = w_i(x, t)$, by solving the spatial part of the wave equation (Helmholtz equation) [10, 17]. Such spatially dependent weights can model the diffraction of sound around the scene geometry.

4.3 MODEL REDUCTION OF NONLINEAR DEFORMATIONS

Up to this point, we have considered model reduction of linear systems (Equation 4.2). Linear systems have an important limitation: they produce very visible artifacts under large deformations (see Figure 4.3). These artifacts can be removed by applying model reduction to the nonlinear equations of a deformable object:

$$M\ddot{u} + D\dot{u} + f_{\text{int}}(u) = f. \qquad (4.5)$$

As detailed in the first part of the course, the nonlinearity in the internal forces $f_{\text{int}}(u)$ arises due to large-deformation strain (*geometric nonlinearity*), and due to nonlinearities in the strain-stress relationship. How to apply model reduction to Equation 4.5? We proceed in the same way as with linear systems; we assume the availability of a basis $U \in \mathbb{R}^{3n \times r}$ (r is basis size) and approximate $u = Uz$, where $z \in \mathbb{R}^r$ is the vector of *reduced coordinates*. After projection by U^T, we obtain

$$\ddot{z} + U^T DU\dot{z} + U^T f_{\text{int}}(Uz) = U^T f. \qquad (4.6)$$

This equation determines the dynamics of the reduced coordinates $z = z(t) \in \mathbb{R}^r$, and therefore also the dynamics of $u(t) = Uz(t)$. Equation 4.5 is similar to Equation 4.3, except that it is nonlinear and the components of z are **not** decoupled. At this point, two questions emerge: (1) How can we timestep Equation 4.6? (2) How do we choose the basis U?

4.3.1 TIMESTEPPING THE REDUCED NONLINEAR EQUATIONS OF MOTION

In order to timestep Equation 4.5, one needs to evaluate the *reduced internal forces*, $\tilde{f}_{\text{int}} = U^T f_{\text{int}}(Uz)$, for arbitrary configurations $z \in \mathbb{R}^r$. Furthermore, as equations of elasticity are typically stiff, implicit integration is required, necessitating a further derivative of $\tilde{f}_{\text{int}}$, the *reduced tangent stiffness matrix*

$$\tilde{K}(z) = \frac{d\tilde{f}_{\text{int}}}{dz} = U^T K(Uz)U \in \mathbb{R}^{r \times r}. \qquad (4.7)$$

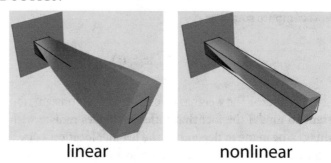

linear   nonlinear

Figure 4.3: Model reduction applied to a linear and nonlinear system.

Here $K(u) = df_{\text{int}}/du$ is the (unreduced) tangent stiffness matrix in configuration u. Note that, in general, the term $\tilde{f}_{\text{int}}$ cannot be algebraically simplified; its evaluation must proceed by first forming Uz, then evaluating $f_{\text{int}}(Uz)$, and finally forming a projection by pre-multiplying with U^T. Evaluation of $\tilde{K}(z)$ is even more complex. Once $\tilde{f}_{\text{int}}(z)$ and $\tilde{K}(z)$ are known, one can use any implicit integrator to timestep the system (see [4] for details). The key important fact is that this integrator will need to solve a dense $r \times r$ linear system as opposed to a sparse $3n \times 3n$ system, as is the case with implicit integration of unreduced systems. Since $r \ll 3n$, this usually leads to significant computational savings.

How to evaluate $\tilde{f}_{\text{int}}(z)$ and $\tilde{K}(z)$ in practice? If the simulation is geometrically nonlinear, but materially *linear*, then it can be shown [4] that each component of $f_{\text{int}}(u)$ is a *cubic polynomial* in the components of u [4]. Consequently, $\tilde{f}_{\text{int}}$ are cubic polynomials in the components of z. Note that this is a manifestation of a more general principle: for any polynomial function $G(u)$, its projection $U^T G(Uz)$ will be a polynomial in z, of the same degree. Treuille and colleagues [39], for example, exploited this fact with quadratic advection forces for reduced fluids. For geometrically nonlinear materials, one can precompute the coefficients of the cubic polynomials. As there are r components of the reduced force, each of which is a cubic polynomial in r variables, the necessary storage is $O(r^4)$. For moderate values of r ($r < 30$), this storage is manageable (under one megabyte; details are in [4]). Because the reduced stiffness matrix $\tilde{K}(z)$ is the gradient of $\tilde{f}_{\text{int}}$ with respect to z, the reduced stiffness matrix is a quadratic function in z with coefficients directly related to those of the reduced internal forces. For exact evaluation of internal forces and tangent stiffness matrices, all polynomial terms must be "touched" exactly once. Therefore, the cost of evaluation of reduced internal forces and tangent stiffness matrices is $O(r^4)$, whereas the cost of implicit integration is $O(r^3)$.

For general materials, An and colleagues [2] have designed a fast approximation scheme that can decrease the reduced internal force and stiffness matrix computation time to $O(r^2)$ and $O(r^3)$, respectively. For simulations that use implicit integration, the runtime complexity is therefore $O(r^3)$. The method works by observing that the elastic strain energy $E(z)$ and internal forces $\tilde{f}_{\text{int}}(z)$, for reduced coordinates z, are obtained by integration of the energy density $\Psi(X, z)$ and its gradient over the entire mesh:

$$E(z) = \int_{\Omega} \Psi(X, z) dV, \qquad (4.8)$$

$$\tilde{f}_{\text{int}}(z) = \int_{\Omega} \frac{\partial \Psi(X, z)}{\partial z} dV. \qquad (4.9)$$

As opposed to evaluating $\tilde{f}_{\text{int}}$ using the exact formula $\tilde{f}_{\text{int}} = U^T f_{\text{int}}(Uz)$, An and colleagues approximate the integral in (4.9) using numerical quadrature. In order to do so, they determine positions $X_i \in \Omega$, and weights $w_i \in \mathbb{R}$, such that

$$\tilde{f}_{\text{int}}(z) = \int_{\Omega} \frac{\partial \Psi(X, z)}{\partial z} dV \approx \sum_{i=1}^{T} w_i \, g(X_i, z), \qquad (4.10)$$

$$\tilde{K}(z) \approx \sum_{i=1}^{T} w_i \frac{\partial g(X_i, z)}{\partial z}, \qquad (4.11)$$

where $g(X, z) = \partial \Psi(X, z)/\partial z$. At runtime, given a value z, one then only has to evaluate $g(X_i, z)$ and $\partial g(X_i, z)/\partial z$, for $i = 1, \ldots, T$ and sum the terms together. The number of quadrature points T is usually set to $T = r$. Positions and weights are obtained using a training process. Given a set of representative "training" reduced coordinates $z^{(1)}, \ldots, z^{(N)}$, the method computes positions and weights that best approximate the reduced force $\tilde{f}_{\text{int}}$ for these training datapoints. To avoid overfitting and to keep the stiffness matrix symmetric positive-definite, the weights w_i are chosen to be non-negative, using nonnegative least squares (NNLS) [25]. The positions X_i are determined using a greedy approach, designed to minimize the NNLS error residual (details in [2]).

4.3.2 CHOICE OF BASIS

The matrix U is a time-invariant matrix specifying a basis of some r-dimensional ($r \ll 3n$) linear subspace of $\mathbb{R}^{3n}$. The basis is assumed to be mass-orthonormal, i.e., $U^T M U = I$. If this is not the case, one can easily convert U to such a basis using a mass-weighted Gramm-Schmidt process. For each fixed $r > 1$, there is an infinite number of possible choices for the linear subspace and for its basis. Good subspaces are low-dimensional spaces that well-approximate the space of typical nonlinear deformations. The choice of subspace depends on geometry, boundary conditions, and material properties. Selection of a good subspace is a non-trivial problem. We now present two choices: basis from simulation data ("POD basis"), and basis from modal derivatives. The former requires pre-simulation (using a general deformable solver), whereas the latter can create a basis automatically without pre-simulation.

Basis from Simulation Data ("POD basis")

In model reduction literature, a very common approach to create a basis for nonlinear systems is to obtain some "snapshots" of the system, $u_1, u_2, \ldots, u_N$, and then use statistical techniques to extract a representative low-dimensional space. The snapshots can be obtained by running a full (unreduced) simulation, or using measurements of a real system. Given the snapshots, one obtains the subspace U by performing singular value decomposition (SVD) on $A = [u_1, u_2, \ldots, u_N]$,

$$A = U \Sigma V^T, \tag{4.12}$$

and retains the columns of U corresponding to the largest r singular values. In practice, it is advantageous to apply SVD not with respect to the standard inner product $y^T x$ in $\mathbb{R}^{3n}$, but with respect to a mass-weighted inner-product $y^T M x$ (*mass-PCA*; see [4] for details).

It is challenging to measure transient volumetric deformation fields with high accuracy [9, 22]. Therefore, large-deformation model reduction applications in computer graphics have so far relied on simulation data, or on modal derivatives, which we describe next.

Modal Derivatives

Linear modal analysis (Section 4.2) provides a quality deformation basis for small deformations away from the rest pose. The linear modes, however, are not a good basis for large deformations, because they lack the deformations that automatically "activate" in a nonlinear system. For example, when a cantilever beam deflects sideways in the direction of the first linear mode, it also simultaneously compresses, in a very specific, non-uniform way. This happens automatically in a nonlinear system. A linear basis, however, lacks the proper (non-uniform) compression mode, and therefore the system projected onto the linear basis will be stiff (it "locks"). In practice, such locking manifests as a rapid loss of energy (numerical damping), and as an increase in the natural oscillation frequencies of the system, a phenomenon also observed with model reduction of electrical circuits [13]. One could attempt to resolve these issues by retaining a larger number of linear modes. Such an approach is, however, not very practical with nonlinear systems, because a very large number of modes would be needed in practice, whereas the time to solve the reduced system for implicit integration scales as $O(r^3)$.

These problems can be remedied using modal derivatives: deformations that naturally co-appear in a nonlinear system when the system is excited in the direction of linear modes. By forming a basis that consists of both linear modes and their modal derivatives, we arrive at a compact, low-dimensional basis, which can represent large deformations and can be computed purely based on the mesh geometry and material properties; no advance knowledge of run-time forcing or pre-simulation is required. We will illustrate modal derivatives for deformable objects that are sufficiently constrained so that they do not possess rigid degrees of freedom, but modal derivatives can also be computed for unconstrained systems. Under a static load f, the system will deform into a deformation u, where u satisfies the unreduced static equation $f_{\text{int}}(u) = f$. Consider what happens if we statically load the system into the direction of linear modes. In particular, suppose we apply a static force load $M U_{\text{lin}} \Lambda p$, where M is the mass matrix, $U_{\text{lin}} = [\psi_1, \psi_2, \ldots, \psi_k]$ is the linear modal matrix, Λ is the diagonal matrix of squared frequencies $\text{diag}(\omega_1^2, \ldots, \omega_k^2)$, and $p \in \mathbb{R}^k$ is some parameter that controls the strength of each mode in the load. It can be easily verified that these are the force loads that, for small deformations, produce deformations

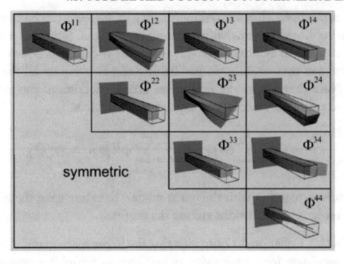

Figure 4.4: Modal derivatives for a cantilever beam.

within the space spanned by the linear modes. Given a p, we can solve the nonlinear equation $f_{\text{int}}(u) = M U_{\text{lin}} \Lambda p$ for u, i.e., we can define a unique function $u = u(p)$ (mapping from $\mathbb{R}^k$ to $\mathbb{R}^{3n}$, and C^∞ differentiable), such that

$$f_{\text{int}}(u(p)) = M U_{\text{lin}} \Lambda p, \qquad (4.13)$$

for every $p \in \mathbb{R}^k$ in some sufficiently small neighborhood of the origin in $\mathbb{R}^k$. Can we compute the Taylor series expansion of u in terms of p? By differentiating Equation 4.13 with respect to p, one obtains

$$\frac{\partial f_{\text{int}}}{\partial u} \frac{\partial u}{\partial p} = M U_{\text{lin}} \Lambda, \qquad (4.14)$$

which is valid for all p in some small neighborhood of the origin of $\mathbb{R}^k$. In particular, for $p = 0^k$, we get $K \frac{\partial u}{\partial p} = M U_{\text{lin}} \Lambda$. Therefore, $\frac{\partial u}{\partial p} = U_{\text{lin}}$, i.e., the first-order responses of the system are the linear modes, as expected. To compute the second order derivatives of u, we differentiate Equation 4.14 one order further by p, which, when we set $p = 0^k$, gives us

$$K \frac{\partial^2 u}{\partial p_i \partial p_j} = -(H : \psi_j)\psi_i. \qquad (4.15)$$

Here, H is the *Hessian stiffness tensor*, the first derivative of the tangent stiffness matrix, evaluated at $u = 0$ (see [4]). The deformation vectors

$$\Phi^{ij} = \frac{\partial^2 u}{\partial p_i \partial p_j} \qquad (4.16)$$

are called *modal derivatives*. They are symmetric, $\Phi_{ij} = \Phi_{ji}$, and can be computed from Equation 4.15 by solving linear systems with a constant matrix K (stiffness matrix of the origin). Because K is constant and symmetric positive-definite, it can be pre-factored using Cholesky factorization. One can then rapidly (in parallel if desired) compute all the modal derivatives, $0 \le i \le j < k$. Note that the modal derivatives are, by definition, the second derivatives of $u = u(p)$. The second-order Taylor series expansion is therefore

$$u(p) = \sum_{i=1}^{k} \Psi^i p_i + \frac{1}{2} \sum_{i=1}^{k} \sum_{j=1}^{k} \Phi^{ij} p_i p_j + O(p^3). \tag{4.17}$$

The modal derivatives, together with the linear modes, therefore span the natural second-order system response for large deformations around the origin.

Creating the basis U.   Equation 4.17 suggests that the linear space spanned by all vectors Ψ^i and Φ^{ij} is a natural candidate for a basis (after mass-Gramm-Schmidt mass-orthonormalization). However, its dimension $k + k(k + 1)/2$ may be prohibitive for real-time systems. In practice, we obtain a smaller basis by scaling the modes and derivatives according to the eigenvalues of the corresponding linear modes and applying mass-PCA on the "dataset"

$$\left\{ \frac{\lambda_1}{\lambda_j} \Psi^j \mid j = 1, \dots, k \right\} \cup \left\{ \frac{\lambda_1^2}{\lambda_i \lambda_j} \Phi^{ij} \mid i \le j; \ i, j = 1, \dots, k \right\}. \tag{4.18}$$

The scaling puts greater weight on dominant low-frequency modes and their derivatives, which could otherwise be masked by high-frequency modes and derivatives.

4.4   MODEL REDUCTION AND DOMAIN DECOMPOSITION

Model reduction as described in the previous section is global: it reduces the entire object using a single, global basis. Unless r is large, it is difficult to capture local detail using such a basis. Because the computation time grows at least as $O(r^3)$ (time to solve the dense $r \times r$ system [2]), large values of r (several hundreds of modes) are not practical. Therefore, it is natural to ask if the object can somehow be decomposed into smaller pieces, each of which is reduced separately, and then the pieces are connected into a global system. This is the idea of *domain decomposition*, a classical technique in applied mathematics and engineering. In engineering applications, however, the deformations are typically small. In computer graphics, we have to accommodate large deformations (e.g., rotations) in the interfaces joining two domains, which means that standard domain decomposition techniques simply cannot be extended to computer graphics problems.

In computer graphics, domain decomposition for deformable models has initially been applied to small domain deformations with running times dependent on the number of domain and interface vertices. For example, a linear quasi-static application using Green's functions has been

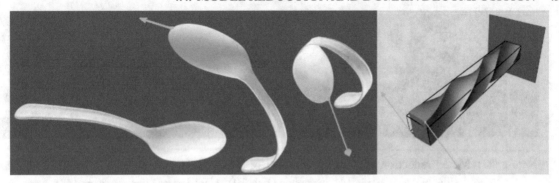

Figure 4.5: Extreme shapes captured by modal derivatives: Although modal derivative are computed about the rest pose, their deformation subspace contains sufficient nonlinear content to describe large deformations. Left: Spoon ($k = 6, r = 15$) is constrained at far end. Right: Beam ($r = 5$, twist angle=270°) is simulated in a subspace spanned by "twist" linear modes and their derivatives $\Psi^4, \Psi^9, \Phi^{44}, \Phi^{49}, \Phi^{99}$.

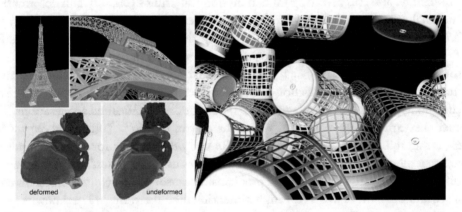

Figure 4.6: Reduced simulations: Left: model reduction enables interactive simulations of nonlinear deformable models. Right: reduction also enables fast large-scale multibody dynamic simulations, with nonlinear deformable objects undergoing free-flight motion. Collisions among the 512 baskets were resolved using BD-Trees [20].

presented in [19], whereas Huang and colleagues [15] exploited redundancy in stiffness matrix inverses to combine linear FEM with domain decomposition. Recently, domain decomposition under large deformations has received significant attention in computer-graphics literature. Barbič and Zhao [8] demonstrated a domain decomposition method by employing gradients of polar decomposition rotation matrices, whereas Kim and James [24] tackled a similar problem using inter-domain spring forces.

Figure 4.7: Model reduction with a large number of localized degrees of freedom: Left: nonlinear reduced simulation of an oak tree (41 branches ($r = 20$), 1394 leaves ($r = 8$), $d = 1435$ domains, $\hat{r} = 11,972$ total DOFs) running at 5 fps. Right: simulation detail.

4.5   MODEL REDUCTION AND CONTROL

Optimal control problems occur frequently in computer animation. Often, they are cast as *space-time optimization problems* involving human motion [34], fluids [30, 40], and deformations [42]. With optimal control of deformable objects, one seeks a sequence of forces (*control vectors*) $f_i \in \mathbb{R}^{3n}$, $i = 0, ..., T - 1$, such that the resulting deformations (*state vectors*) $u_i \in \mathbb{R}^{3n}$, $i = 0, ..., T - 1$, obtained by timestepping Equation 4.5 forward in time under those forces, minimize some scalar objective $E(u_0, ..., u_{T-1})$. The scalar objective typically includes terms such as magnitude of control vectors, deviation from some reference trajectory, deviation from keyframes at specific moments of time, and magnitude of deformation velocities and accelerations. The forces are sometimes expressed as $f_i = Bg_i$, where the matrix $B \in \mathbb{R}^{3n \times m}$ gives the *control basis*. The problem is said to be *underactuated* when $m < 3n$ and *fully actuated* for $m = 3n$. Underactuated problems can model objects that can propel themselves using "muscles," and are generally much more difficult to solve than fully actuated problems. Because optimal control problems compound both space and time, they have a very high dimensionality: there are $3nT$ unknowns (the control vectors f_i) in the optimal control problem. Such a huge state space leads to optimization problems that diverge, converge to local minima, or take a very long time to converge to a plausible solution.

Model reduction is very beneficial to optimal control because it greatly reduces the state and control size. The states u_i are replaced with the reduced states z_i, the control vectors f_i are replaced with the reduced internal forces $\tilde{f}_i$, and Equation 4.5 is replaced with its reduced version (Equation 4.6). Although such a reduced optimal control problem only approximates the original problem, its dimensionality is only $rT \ll 3nT$; therefore, the occurrence of local minima is greatly decreased. Optimal reduced forces can be found faster than unreduced forces, because one can rapidly explore the solution space by running many reduced forward simulations and by quickly evaluating the reduced objective gradient [5] (Figure 4.8). Standard controllers such as the linear-quadratic regulator [37] are impractical with deformable objects as they involve dense $3n \times 3n$

Figure 4.8: Fast authoring of animations with dynamics [5]: This soft-body dinosaur sequence consists of five walking steps, and includes dynamic deformation effects due to inertia and impact forces. Each step was generated by solving a space-time optimization problem, involving three user-provided keyframes, and requiring only three minutes total to solve due to a proper application of model reduction to the control problem. Unreduced optimization took one hour for each step. The four images show output poses at times corresponding to four consecutive keyframes (out of 11 total). For comparison, the keyframe is shown in the top-left of each image.

gain matrices. With reduction, however, such control becomes feasible as the gain matrices are now much smaller ($r \times r$). Barbič and Popović [7] exploited such a combination of LQR control and model reduction for real-time tracking of nonlinear deformable object simulations, using minimal ("gentle") forces.

4.6   FREE SOFTWARE FOR MODEL REDUCTION

The implementation of [6] (by Jernej Barbič) is freely available on the web (BSD license). It is a part of Vega FEM 2.0, a general-purpose free simulator for FEM nonlinear 3D deformable objects (including deformable dynamics). It can be downloaded from: `http://www.jernejba rbic.com/vega`. It includes

1. a precomputation utility to compute linear modes (Equation 4.1) and modal derivatives (Equation 4.15), and to construct the simulation basis U (mass-PCA applied to the dataset of Equation 4.18); optionally, the basis can also be computed from pre-existing simulation data ("POD basis");

2. a precomputation utility to compute the cubic polynomial coefficients for reduced internal forces $\tilde{f}_{\text{int}}$ and stiffness matrices $\tilde{K}$ (Section 4.3.1), for isotropic geometrically nonlinear material models (*St. Venant-Kirchhoff material*); and

3. an efficient C/C++ library to timestep the reduced model precomputed in the above steps 1, 2 (Equation 4.6), and an example run-time driver.

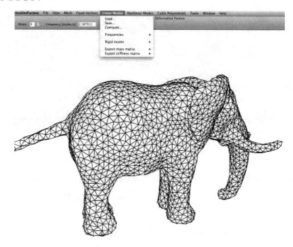

Figure 4.9: Precomputation utility to compute linear modes, derivatives, basis for large deformations, and cubic polynomial coefficients. Free (BSD license), available at http://www.jernejbarbic.com /vega.

4.7   DEFORMATION WARPING

As outlined in Section 4.3, linear modal analysis (Section 4.2) leads to visible artifacts under large deformations, and these artifacts can be removed by applying model reduction to the nonlinear equations of motion. Deformation warping is an alternative, purely geometric, approach to remedy the same problem. The idea is to keep Equation 4.2 as the underlying dynamic equation, but to post-process the resulting deformations u using a geometric "filter" that removes large-deformation artifacts. For example, given a tetrahedral mesh, warping establishes a mapping that maps linearized deformations $u \in \mathbb{R}^{3n}$ (away from the rest configuration) to "good-looking" large deformations $q = W(u) \in \mathbb{R}^{3n}$ (also away from the rest configuration). The user is only shown the corrected deformations q. Warping is very robust. For example, twisting deformations, with local rotations as large as several complete 360 degrees cycles, can be easily accommodated. The underlying dynamics, however, is still linear, and this is visible with large-deformation motion as linear deformations essentially follow a sinusoidal curve, $\sin(\omega_i t)$. In contrast, objects simulated using nonlinear methods usually stiffen under large deformations, and therefore spend a smaller percentage of the oscillation cycle time at large deformations.

The idea that modes could be warped to correct large deformation artifacts was first observed by Choi and Ko [11]. They noticed that, by taking the curl of each modal vector, one can derive per-vertex infinitesimal rotations due to the activation of each mode. These rotations can then be integrated in time, resulting in large deformations free of artifacts. Although not surveyed here in detail, their approach is fast, and laid the foundation for other warping methods developed later.

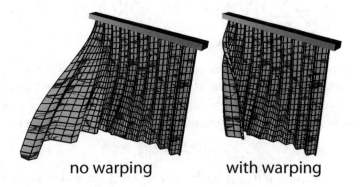

no warping   with warping

Figure 4.10: Warping corrects linearization artifacts under large deformations.

4.7.1   ROTATION-STRAIN COORDINATE WARPING

In these notes, we describe a recent efficient flavor of warping, the *rotation-strain coordinate warping* [16]. Let $G_j \in \mathbb{R}^{9 \times 3n}$ be the discrete gradient operator of tet j, i.e., deformation gradient of tet j under deformation $u \in \mathbb{R}^{3n}$ equals $I + G_j u$. Decompose the 3×3 matrix $G_j u$ into a symmetric and antisymmetric component,

$$G_j u = \frac{G_j u + (G_j u)^T}{2} + \frac{G_j u - (G_j u)^T}{2}. \qquad (4.19)$$

We can then denote the upper triangle of the symmetric part as $\epsilon_j \in \mathbb{R}^6$, and the skew-vector corresponding to the antisymmetric part as $\omega_j \in \mathbb{R}^3$. We then assemble $[\epsilon_j, \omega_j]$ for all tets into a vector $y(u) \in \mathbb{R}^{9 \#tets}$; denote $y_{\epsilon,j} = \epsilon_j$ and $y_{\omega,j} = \omega_j$. Vector y forms the *rotation-strain coordinates* of the mesh. Given a linearized deformation u, we then postulate that we should seek a deformation q so that the deformation gradient inside tet j equals

$$\exp(\widetilde{y_{\omega,j}})(I + \text{sym}(y_{\epsilon,j})). \qquad (4.20)$$

Here, $\text{sym}(x)$ is the 3×3 symmetric matrix corresponding to its upper-triangle $x \in \mathbb{R}^6$, $\tilde{x}$ is the 3×3 skew-symmetric cross-product matrix corresponding to vector $x \in \mathbb{R}^3$ ($\tilde{x}v = x \times v$ for all $v \in \mathbb{R}^3$), and exp is the matrix exponential function [36]. This condition cannot be satisfied for all the tets simultaneously. Instead, given u, we find a deformation q under which the deformation gradients are as close as possible to those given by Equation 4.20. This can be done by solving the following least-squares problem:

$$\underset{q}{\mathrm{argmin}} \sum_{j=1}^{\#tets} V_j || I + G_j q - \exp(\widetilde{y_{\omega,j}})(I + \mathrm{sym}(y_{\epsilon,j})) ||_F^2, \tag{4.21}$$

$$\text{subject to pinned vertices} \tag{4.22}$$

where $|| \; ||_F$ denotes the Frobenius norm of a 3×3 matrix, and V_j is the volume of tet j. The pinned vertices are the vertices where the model is rooted to the ground (boundary conditions). For free-flying objects, a constraint can be formed that keeps the center of mass unmodified. The objective function in Equation 4.21 is quadratic in q, and can be rewritten as

$$|| VGq - b ||_2^2, \tag{4.23}$$

for $V = \mathrm{diag}(\sqrt{V_1}, \sqrt{V_1}, \ldots, \sqrt{V_{\#tets}})$ (each entry repeated 9x), and where $G \in \mathbb{R}^{9\#tets \times 3n}$ is the gradient matrix assembled from all G_j. The nine-block of vector $b \in \mathbb{R}^{9\#tets}$ corresponding to tet j, expressed as a row-major 3×3 matrix, equals

$$b_j = \sqrt{V_j}\left(\exp(\widetilde{y_{\omega,j}})(I + \mathrm{sym}(y_{\epsilon,j})) - I\right). \tag{4.24}$$

In a typical tet mesh, there are more tets than vertices, therefore, the optimization problem is overconstrained. The minimization can be performed via Lagrange multipliers, by solving

$$\begin{bmatrix} L & d \\ d^T & 0 \end{bmatrix} \begin{bmatrix} q \\ \lambda \end{bmatrix} = \begin{bmatrix} (VG)^T b \\ 0 \end{bmatrix}, \tag{4.25}$$

where $L = G^T V^2 G$, and where d corresponds to the pinned input vertices (note: an implementation can simply remove the rows-columns from L; this is equivalent). Matrix L is called the *discrete Laplacian of the mesh*. It only depends on the input mesh geometry, and not on U or u. The system matrix in Equation 4.25 is sparse and constant, and can be pre-factored, so warping can be performed efficiently at runtime.

4.7.2   WARPING FOR TRIANGLE MESHES

Triangle meshes are commonly employed in computer graphics, say, for simulation of thin shells and cloth. Such physical systems also produce the mass matrix M and stiffness matrix K. Therefore, the small deformation analysis (Equation 4.2) and model reduction (as in Equation 4.3) apply also to such problems. In order to apply warping, however, we must define deformation gradients for each triangle. As observed by Sumner and Popović [38], the three vertices of a triangle before and after deformation do not fully determine the affine transformation since they do not establish how the space perpendicular to the triangle deforms. They resolve this issue by adding a (fictitious) fourth vertex v_4,

$$v_4 = v_1 + \frac{(v_2 - v_1) \times (v_3 - v_1)}{\sqrt{|(v_2 - v_1) \times (v_3 - v_1)|}}, \tag{4.26}$$

where v_1, v_2, v_3 are the triangle vertices. Vertices v_1, v_2, v_3, v_4 define a tetrahedron. Let v_1', v_2', v_3' denote the deformed vertex positions; then, we can use (4.26) to compute the deformed fictitious vertex v_4'. The deformation gradient F for the triangle equals

$$
F = \left[\begin{array}{ccc} v_2' - v_1' & v_3' - v_1' & v_4' - v_1' \end{array} \right] \left[\begin{array}{ccc} v_2 - v_1 & v_3 - v_1 & v_4 - v_1 \end{array} \right]^{-1} . \tag{4.27}
$$

Given the deformation gradient, warping then proceeds in the same way as described for tetrahedral meshes in previous sections. A more principled version of triangle mesh warping has been presented by [12].

4.8 ACKNOWLEDGMENTS

This work was sponsored by the National Science Foundation (CAREER-53-4509-6600). I would like to thank Eftychios Sifakis and Yili Zhao for helpful suggestions.

SUMMARY

In this chapter we have discussed model reduction and its applications to large deformation modeling and animation in computer graphics. We explained how small-deformation analysis can be extended to the large deformation regime and surveyed several key research papers in this area.

Bibliography

[1] V. Akcelik, J. Bielak, G. Biros, I. Epanomeritakis, A. Fernandez, O. Ghattas, E. J. Kim, J. Lopez, D. O'Hallaron, T. Tu, and J. Urbanic. High-resolution forward and inverse earthquake modeling on terascale computers. In *Proceedings of ACM/IEEE SC2003*, 2003. DOI: 10.1145/1048935.1050202. 38

[2] S. S. An, T. Kim, and D. L. James. Optimizing cubature for efficient integration of subspace deformations. *ACM Trans. on Graphics*, 27(5):165:1–165:10, 2008. DOI: 10.1145/1409060.1409118. 36, 40, 41, 44

[3] D. Baraff and A. P. Witkin. Large steps in cloth simulation. In *Proc. of ACM SIGGRAPH 98*, pages 43–54, July 1998. DOI: 10.1145/280814.280821. 37

[4] J. Barbič. *Real-time Reduced Large-Deformation Models and Distributed Contact for Computer Graphics and Haptics*. Ph.D. thesis, Carnegie Mellon University, Aug. 2007. 36, 40, 42, 43

[5] J. Barbič, M. da Silva, and J. Popović. Deformable object animation using reduced optimal control. *ACM Trans. on Graphics (SIGGRAPH 2009)*, 28(3):53:1–53:9, 2009. DOI: 10.1145/1576246.1531359. 46, 47

[6] J. Barbič and D. L. James. Real-time subspace integration for St. Venant-Kirchhoff deformable models. *ACM Trans. on Graphics*, 24(3):982–990, 2005. DOI: 10.1145/1073204.1073300. 36, 47

[7] J. Barbič and J. Popović. Real-time control of physically based simulations using gentle forces. *ACM Trans. on Graphics (SIGGRAPH Asia 2008)*, 27(5):163:1–163:10, 2008. DOI: 10.1145/1457515.1409116. 47

[8] J. Barbič and Y. Zhao. Real-time large-deformation substructuring. *ACM Trans. on Graphics (SIGGRAPH 2011)*, 30(4):91:1–91:7, 2011. DOI: 10.1145/2010324.1964986. 36, 45

[9] B. Bickel, M. Baecher, M. Otaduy, W. Matusik, H. Pfister, and M. Gross. Capture and modeling of non-linear heterogeneous soft tissue. *ACM Trans. on Graphics (SIGGRAPH 2009)*, 28(3):89:1–89:9, 2009. DOI: 10.1145/1576246.1531395. 42

[10] J. N. Chadwick, S. S. An, and D. L. James. Harmonic Shells: A practical nonlinear sound model for near-rigid thin shells. *ACM Transactions on Graphics*, 28(5):1–10, 2009. DOI: 10.1145/1618452.1618465. 36, 38, 39

[11] M. G. Choi and H.-S. Ko. Modal Warping: Real-time simulation of large rotational deformation and manipulation. *IEEE Trans. on Vis. and Comp. Graphics*, 11(1):91–101, 2005. DOI: 10.1109/TVCG.2005.13. 48

[12] M. G. Choi, S. Y. Woo, and H.-S. Ko. Real-time simulation of thin shells. *Eurographics 2007*, pages 349–354, 2007. DOI: 10.1111/j.1467-8659.2007.01057.x. 51

[13] L. Daniel. Private correspondence with Prof. Luca Daniel, MIT. 42

[14] K. K. Hauser, C. Shen, and J. F. O'Brien. Interactive deformation using modal analysis with constraints. In *Proc. of Graphics Interface*, pages 247–256, 2003. 36

[15] J. Huang, X. Liu, H. Bao, B. Guo, and H.-Y. Shum. An efficient large deformation method using domain decomposition. *Computers & Graphics*, 30(6):927–935, 2006. DOI: 10.1016/j.cag.2006.08.014. 45

[16] J. Huang, Y. Tong, K. Zhou, H. Bao, and M. Desbrun. Interactive shape interpolation through controllable dynamic deformation. *IEEE Trans. on Visualization and Computer Graphics*, 17(7):983–992, 2011. DOI: 10.1109/TVCG.2010.109. 49

[17] D. L. James, J. Barbič, and D. K. Pai. Precomputed acoustic transfer: Output-sensitive, accurate sound generation for geometrically complex vibration sources. *ACM Transactions on Graphics (SIGGRAPH 2006)*, 25(3), 2006. DOI: 10.1145/1179352.1141983. 36, 38, 39

[18] D. L. James and D. K. Pai. DyRT: Dynamic response textures for real-time deformation simulation with graphics hardware. *ACM Trans. on Graphics*, 21(3):582–585, 2002. DOI: 10.1145/566654.566621. 36, 38

[19] D. L. James and D. K. Pai. Real-time simulation of multizone elastokinematic models. In *IEEE Int. Conf. on Robotics and Automation*, pages 927–932, 2002. DOI: 10.1109/ROBOT.2002.1013475. 45

[20] D. L. James and D. K. Pai. BD-Tree: Output-sensitive collision detection for reduced deformable models. *ACM Trans. on Graphics*, 23(3):393–398, 2004. DOI: 10.1145/1015706.1015735. 45

[21] D. M. Kaufman, S. Sueda, D. L. James, and D. K. Pai. Staggered projections for frictional contact in multibody systems. *ACM Transactions on Graphics*, 27(5):164:1–164:11, 2008. DOI: 10.1145/1409060.1409117. 36

[22] A. E. Kerdok, S. M. Cotin, M. P. Ottensmeyer, A. M. Galea, R. D. Howe, and S. L. Dawson. Truth cube: Establishing physical standards for soft tissue simulation. *Medical Image Analysis*, 7(3):283–291, 2003. DOI: 10.1016/S1361-8415(03)00008-2. 42

[23] T. Kim and D. James. Skipping steps in deformable simulation with online model reduction. *ACM Trans. on Graphics (SIGGRAPH Asia 2009)*, 28(5):123:1–123:9, 2009. DOI: 10.1145/1666611.1666627. 36

[24] T. Kim and D. James. Physics-based character skinning using multi-domain subspace deformations. In *Symp. on Computer Animation (SCA)*, pages 63–72, 2011. DOI: 10.1145/2019406.2019415. 36, 45

[25] C. L. Lawson and R. J. Hanson. *Solving Least Square Problems*. Prentice Hall, Englewood Cliffs, NJ, 1974. 41

[26] R. Lehoucq, D. Sorensen, and C. Yang. ARPACK Users' Guide: Solution of large-scale eigenvalue problems with implicitly restarted Arnoldi methods. Technical report, Comp. and Applied Mathematics, Rice Univ., 1997. DOI: 10.1137/1.9780898719628. 37

[27] J.-R. Li. *Model Reduction of Large Linear Systems via Low Rank System Gramians*. Ph.D. thesis, Massachusetts Institute of Technology, 2000. 35

[28] R.-C. Li and Z. Bai. Structure preserving model reduction using a Krylov subspace projection formulation. *Comm. Math. Sci.*, 3(2):179–199, 2005. DOI: 10.4310/CMS.2005.v3.n2.a6. 35

[29] J. L. Lumley. The structure of inhomogeneous turbulence. In A.M.Yaglom and V.I.Tatarski, editors, *Atmospheric turbulence and wave propagation*, pages 166–178, 1967. 35

[30] A. McNamara, A. Treuille, Z. Popović, and J. Stam. Fluid control using the adjoint method. *ACM Trans. on Graphics (SIGGRAPH 2004)*, 23(3):449–456, 2004. DOI: 10.1145/1015706.1015744. 46

[31] J. F. O'Brien, C. Shen, and C. M. Gatchalian. Synthesizing sounds from rigid-body simulations. In *Symp. on Computer Animation (SCA)*, pages 175–181, 2002. DOI: 10.1145/545261.545290. 38, 39

[32] A. Pentland and J. Williams. Good vibrations: Modal dynamics for graphics and animation. *Computer Graphics (Proc. of ACM SIGGRAPH 89)*, 23(3):215–222, 1989. DOI: 10.1145/74334.74355. 36

[33] M. Rewienski. *A Trajectory Piecewise-Linear Approach to Model Order Reduction of Nonlinear Dynamical Systems*. Ph.D. thesis, Massachusetts Institute of Technology, 2003. 35

[34] A. Safonova, J. Hodgins, and N. Pollard. Synthesizing physically realistic human motion in low-dimensional, behavior-specific spaces. *ACM Trans. on Graphics (SIGGRAPH 2004)*, 23(3):514–521, 2004. DOI: 10.1145/1015706.1015754. 46

[35] A. A. Shabana. *Theory of Vibration, Volume II: Discrete and Continuous Systems.* Springer–Verlag, New York, NY, 1990. DOI: 10.1007/978-1-4684-0380-0. 37

[36] R. B. Sidje. Expokit: A software package for computing matrix exponentials. *ACM Trans. on Mathematical Software*, 24(1):130–156, 1998. www.expokit.org. DOI: 10.1145/285861.285868. 49

[37] R. F. Stengel. *Optimal Control and Estimation.* Dover Publications, New York, 1994. 46

[38] R. Sumner and J. Popović. Deformation transfer for triangle meshes. *ACM Trans. on Graphics (SIGGRAPH 2004)*, 23(3):399–405, 2004. DOI: 10.1145/1015706.1015736. 50

[39] A. Treuille, A. Lewis, and Z. Popović. Model reduction for real-time fluids. *ACM Trans. on Graphics*, 25(3):826–834, 2006. DOI: 10.1145/1141911.1141962. 36, 40

[40] A. Treuille, A. McNamara, Z. Popović, and J. Stam. Keyframe control of smoke simulations. *ACM Trans. on Graphics (SIGGRAPH 2003)*, 22(3):716–723, 2003. DOI: 10.1145/882262.882337. 46

[41] M. Wicke, M. Stanton, and A. Treuille. Modular bases for fluid dynamics. *ACM Trans. on Graphics*, 28(3):39:1–39:8, 2009. DOI: 10.1145/1531326.1531345. 36

[42] C. Wojtan, P. J. Mucha, and G. Turk. Keyframe control of complex particle systems using the adjoint method. In *Symp. on Computer Animation (SCA)*, pages 15–23, Sept. 2006. DOI: 10.1145/1218064.1218067. 46

Authors' Biographies

EFTYCHIOS SIFAKIS

Eftychios Sifakis is an Assistant Professor of Computer Sciences and (by courtesy) Mechanical Engineering and Mathematics at the University of Wisconsin-Madison. He obtained his Ph.D. degree in Computer Science (2007) from Stanford University. Between 2007–2010 he was a postdoctoral researcher in the University of California, Los Angeles, with a joint appointment in Computer Science and Mathematics. His research focuses on scientific computing, physics-based modeling and computer graphics. He is particularly interested in biomechanical modeling for applications such as character animation, medical simulations, and virtual surgical environments. Eftychios has served as a research consultant with Intel Corporation, Walt Disney Animation Studios, and SimQuest LLC, and is a co-founder of the Wisconsin Applied Computing Center and recipient of the NSF CAREER award.

JERNEJ BARBIČ

Jernej Barbič is an associate professor of computer science at University of Southern California. In 2011, *MIT Technology Review* named him one of the Top 35 Innovators under the age of 35 in the world (TR35). Jernej's research interests include nonlinear solid deformation modeling, model reduction, collision detection and contact, optimal control, and interactive design of deformations and animations. He is the author of *Vega FEM,* an efficient free C/C++ software physics library for deformable object simulation. Jernej is a Sloan Fellow (2014).

Printed in the United States
by Baker & Taylor Publisher Services